T0213977

SpringerBriefs in Applied Sciences and Technology

SpringerBriefs present concise summaries of cutting-edge research and practical applications across a wide spectrum of fields. Featuring compact volumes of 50 to 125 pages, the series covers a range of content from professional to academic.

Typical publications can be:

- A timely report of state-of-the art methods
- An introduction to or a manual for the application of mathematical or computer techniques
- A bridge between new research results, as published in journal articles
- A snapshot of a hot or emerging topic
- An in-depth case study
- A presentation of core concepts that students must understand in order to make independent contributions

SpringerBriefs are characterized by fast, global electronic dissemination, standard publishing contracts, standardized manuscript preparation and formatting guidelines, and expedited production schedules.

On the one hand, **SpringerBriefs in Applied Sciences and Technology** are devoted to the publication of fundamentals and applications within the different classical engineering disciplines as well as in interdisciplinary fields that recently emerged between these areas. On the other hand, as the boundary separating fundamental research and applied technology is more and more dissolving, this series is particularly open to trans-disciplinary topics between fundamental science and engineering.

Indexed by EI-Compendex, SCOPUS and Springerlink.

More information about this series at http://www.springer.com/series/8884

Peter Hippe

Windup in Control Owing to Sensor Saturation

 Springer

Peter Hippe
Lehrstuhl für Regelungstechnik
Universität Erlangen-Nürnberg
Erlangen, Bayern, Germany

ISSN 2191-530X   ISSN 2191-5318   (electronic)
SpringerBriefs in Applied Sciences and Technology
ISBN 978-3-030-73132-8   ISBN 978-3-030-73133-5   (eBook)
https://doi.org/10.1007/978-3-030-73133-5

This Springer imprint is published by the registered company Springer Nature Switzerland AG
The registered company address is: Gewerbestrasse 11, 6330 Cham, Switzerland

To Helga

Preface

The notion of windup is usually related to input constraints, *i.e.*, to saturating actuators. Due to the limited energy the control signal generated by the linear compensator can only be transmitted by the actuator within a limited amplitude range. And if the compensator contains integral action, the integrating part can attain enormous amplitudes during the saturation period, *i.e.*, it winds up. The undesired effects caused by the integral windup are not the only noxious effects of input constraints. Even in loops with constant state feedback input saturation can have a destabilizing effect. Therefore, it makes sense to differentiate between two types of windup, namely, the *Controller Windup* (caused by the dynamics of the compensator) and the *Plant Windup* (caused by the dynamics of the plant due to a left shift of its eigenvalues by constant state feedback). Controller windup can be prevented by a stabilization of the compensator during input saturation. If this is done by the so-called *Observer Technique*, controller windup is completely prevented, because any remaining undesired effects of input saturation can now be attributed to plant windup. For the prevention of plant windup, one can use an *Additional Dynamic Element* (ADE). This ADE modifies the linear part of the nonlinear loop such that it satisfies one of the criteria for the stability of loops consisting of a linear part and an isolated nonlinear element of the sector type.

Windup prevention caused by input constraints is a well-researched field and there is a vast amount of literature dedicated to it. An overview on the existing methods can be found in [1–13] and in the references therein.

Not only the amplitude range of the actuator, but also the output range of the sensor is limited. Normally the sensors used in practical applications reproduce all emerging amplitudes correctly. However, there are applications where the plant output can be driven past the linear range of the sensor. Thus, if reference signals or disturbances drive the system output to values beyond this range, stability problems, akin to the windup problems associated with input constraints, can occur. This may be caused or exacerbated by integral action in the controller. Therefore, it seems natural to call the undesired effects of sensor saturation also windup.

The reference behavior of constrained systems can always be realized such that neither input nor output saturation cause windup effects. Thus, the real problem in anti-windup control is persistent disturbances causing the saturation elements to become active. Consequently, the predominant part of this contribution is concerned with the attenuation of the effects of step-like or constant disturbances.

Given the vast collection of methods for anti-windup control in the presence of input constraints it looks like an easy task to solve the problem of windup prevention in the presence of output constraints. The literature available, however, tells another story. There are some papers discussing the problem of control and filtering in the presence of sensor saturation but very few investigate the problem of windup prevention in the presence of integral controller action in loops with output-constrained systems.

There exist papers on various aspects of output saturation, such as the stabilization of a system starting from initial conditions [14–16] or the stabilization of the reference behavior [17]. Also the rejection of disturbances has been considered in [18], but only for L_2 bounded signals. An extension to linear discrete-time systems has been presented in [19]. The observability of linear systems with output constraints has been discussed in [20]. For discrete-time systems with sensor saturation, a robust filter was designed in [21] and then applied to digital transmultiplexer systems. In [22], the set-membership filtering problem for discrete-time systems subject to sensor saturation is considered.

Anti-windup control for output-constrained systems is considered in the following papers. In [23], persistent disturbances were allowed, but the scheme does not assure Linear Performance Recovery (LPR) and it leads to eigenvalues of the closed loop whose real parts tend to minus infinity. In [24], various possible configurations of anti-windup systems are discussed, but the authors state that there is no agreed architecture for applying anti-windup to systems with sensor saturation so far. The above two papers are posed in the anti-windup framework, but they do not concentrate on LPR after a sensor saturation event. Furthermore, the solutions rely on an observer estimating the unconstrained output *in addition* to the introduction of an anti-windup compensator.

In contrast, recent work by [25] presents a solution to the sensor saturation problem where LPR is guaranteed. This approach is based on ideas from the hybrid systems literature and has a different architecture from the "essentially linear" compensators presented in other papers. Despite its appealing properties, the compensator is somewhat complicated to implement. Another rather different approach has been recently proposed by [26] where output constraints are interpreted, via a transformation, as constraints on plant inputs and then an (input) anti-windup compensator may be used to prevent instability. This approach is attractive because it will prevent sensor saturation from occurring, but it is potentially conservative due to this.

The reason why a general solution of the anti-windup problem in case of sensor saturation did not exist to date may be due to the fact that it seems so obvious to carry the known methods for input constraints over to the realm of output saturation. And at first glance this seems so natural. If input saturation is active the loop is open so that integral action can increase uncontrolled. If the output signal saturates, the same happens.

Taking a closer look at the problem reveals that output saturation differs in two essential aspects from input saturation. The first relates to the availability of the signals. In input-constrained systems, the unlimited control signal is known and the limited one can easily be obtained by a model of the saturation. In the case of a limited sensor, only the saturated signal is known and if there was a means to also obtain the unlimited one, the whole problem would not exist. The second relates to the measures for windup prevention. In the input-constrained case, the whole range of the limited input signal is available for disturbance rejection also when the compensator is stabilized. This is not true in the output-constrained case. If the set point of the control is close to the sensor saturation limit and persistent disturbances drive the plant output beyond this limit, the input to the compensator can become very small. And this small input signal may then not be large enough to generate a control signal in the bibo-stable controller to bring the plant output back into the linear range of the sensor. As a consequence, LPR would not be guaranteed. A safe rejection of persistent disturbances can only be guaranteed if the integral action remains active during sensor saturation.

Consequently, two methods for anti-windup control in systems with input saturation cannot be carried over to the case of output saturation. The first, namely, *stabilize the compensator during saturation* is not a good idea in view of rejecting persistent disturbances and the second, *modify the dynamics of the linear part during saturation by feeding back the difference between the limited and the unlimited signal* is simply not feasible because the required information is missing.

This contribution presents methods to get around the abovementioned difficulties. Anti-windup structures are presented for Single-Input, Single-Output (SISO) and Multiple-Input, Multiple-Output (MIMO) systems assuring stability for output-constrained systems subject to reference signals and to persistent disturbances. Also presented are methods to prevent windup in the presence of joint sensor and actuator saturation.

Reference inputs can always be applied without causing windup effects in constrained systems. This can either be achieved by an appropriate trajectory planning or by using the nonlinear model-based reference shaping filter originally presented in [6]. This filter shapes arbitrary external reference inputs such that neither input nor output constraints cause windup effects. To make this presentation self-contained, the design of the reference shaping filter is described in the appendix. Also a new technique for reference tracking is presented in the appendix. It can be applied when the controller is designed such that it compensates the plant dynamics. This new method has a much simpler structure (especially for MIMO systems) than the reference filter originally proposed in [6] and it also has the great advantage that it allows a diagonally decoupled reference behavior even for MIMO systems which are not diagonally decouplable by state feedback.

The majority of the results in this contribution suffer from a seeming drawback because they do not follow the classic design paradigm of anti-windup control, namely, "Design an arbitrary linear control and if saturation causes stability problems add appropriate measures to assure stability." Only the method presented in Chap. 1 does so. It uses a saturation indicator to modify the transfer behavior of the compensator in case of sensor saturation such that the loop is stabilized. This method works perfectly, but it has two flaws. It lacks a rigorous proof of stability and it can only be applied to SISO systems.

The main difficulty in proving the stability of the loop with saturation indicator is the series connection of a limiter and a switching element. The outcome of all criteria for the stability of nonlinear systems is the information that the solution presented in Chapter 1 exhibits an unstable behavior. That this unstable behavior does not occur in practice is due to a number of special features, such as coordinated amplitudes of the nonlinear elements and a special design of the linear system driven by them. If one of these features is modified the scheme becomes indeed unstable.

The stability behavior of the loop with saturation indicator is the same as that of a (strictly proven) windup prevention scheme using a not realizable ADE. This allows to solve also the much more complicated problem of joint input and output saturation with the aid of the saturation indicator in Chapter 2. However, when both the output and the input are constrained the classic design paradigm can no longer be applied. Because of the lack of information about the unlimited output, the windup problems triggered by an arbitrarily designed controller can no longer be prevented in a loop with restricted input and output.

The saturation indicator circumnavigates some of the problems resulting from the missing information about the difference between limited and unlimited output signal. And since one cannot obtain precise information about this difference in a loop subject to unknown external disturbances, an anti-windup design with proven stability following the above-cited design paradigm may never be found.

Is it really necessary to stick to the classic design paradigm? The above arguments have shown that the problem of output constraints is quite different from the one of input constraints. And this is also true in view of security considerations, because disturbances can drive the system output to enormous amplitudes during sensor saturation, something not only undesirable in automotive applications. Consequently, one should rethink the whole procedure of windup prevention in the presence of output constraints.

Why not take into account the output constraints right from the start? If the compensator had the property of guaranteeing a stable disturbance behavior in spite of sensor saturation, this would make the additional switching element superfluous so that a rigorous proof was easily feasible. The attenuation of persistent disturbances is an important issue in control. In view of the security reasons mentioned above, it may be desirable that only extraordinary disturbance events drive the output into saturation. Good disturbance attenuation, however, requires fast dynamics and this in turn leads to an intensification of the windup effects.

The solution to this predicament is the name-giving principle of automatic control, namely, compensation. When compensating the plant dynamics by the controller and shifting the remaining poles left, a nearly arbitrary degree of disturbance attenuation can be achieved without causing windup effects. The anti-windup control presented in Chapter 3 uses this compensating approach. It is applicable to SISO and MIMO systems alike and it allows a strict proof of stability. Due to the compensating approach the MIMO version is nearly as simple as the SISO one. It also facilitates the anti-windup design in the presence of input and output saturation.

In the last two chapters, the problem of joint input and output constraints is solved for SISO systems (Chapter 4) and MIMO systems (Chapter 5). Compared to the solution presented in Chapter 2, it has not only the advantage of a strict proof but it also simplifies the design of the stabilizing control in two aspects. Compared to the compensating approach, the dimensions of the design equations in Chapter 2 are blown up due to the substitution of the saturation indicator by the equivalent ADE. And the second simplification arises from the parametrization of the nominal compensator by one parameter only (also in the MIMO case).

The problem with saturation elements at the input *and* the output is, of course, much more involved than that of input *or* output saturation. At the one hand, the stability test is a MIMO problem already for SISO systems and at the other hand, the success of the design process depends on the type of system. For systems with big overshoots, it can happen that no stabilizing control exists. This is true both for SISO and MIMO systems.

With MIMO systems an additional problem occurs, namely, that of deadlock. If external disturbances drive both the inputs and the outputs into saturation it can happen that all or some of the input and output signals remain saturated even after the disturbance inputs have vanished. The tendency for deadlock is a property of the system and it is characterized by its static behavior. There are systems with a strong tendency for deadlock and a slight tendency for it. If a strong tendency for deadlock exists a stabilizing control cannot be designed. For systems with a slight tendency for deadlock, one can usually find a stabilizing control, but the achievable dynamics are often so restricted that they do not warrant practical application. Therefore, the problem of joint input and output constraints should only be attacked for MIMO systems without tendency for deadlock.

In Chapter 5, a criterion is presented whether a system has the tendency for deadlock or not. This simple test is only sufficient for systems with $p < 4$ inputs and outputs. However, the problem of guaranteeing the stability of a loop, where a system with more than three inputs and outputs is coupled to a compensator with integral action, while persistent external disturbances drive all inputs and outputs into saturation for an indefinite interval of time, is a non-trivial problem by all means.

Although not all results in this contribution are proven rigorously, the author hopes that the available results will be helpful and stimulate further research.

Erlangen, Germany   Peter Hippe
March 2021

References

1. D. Bernstein, A. Michel, A chronological bibliography on saturating actuators. Int. J. Robust Nonlinear Control **5**, 375–380 (1995).
2. C. Gökçek, P.T. Kabamba, S.M. Meerkov, An LQR/LQG theory for systems with saturating actuators. IEEE Trans. Autom. Control **46**, 1529–1542 (2001).
3. A. Saberi, Z. Lin, A.R. Teel, Control of linear systems with saturating actuators. IEEE Trans. Autom. Control **41**, 368–378 (1996).

4.   M.C. Turner, L. Zaccarian, Special issue on anti-windup. Int. J. Syst. Sci. **37** (2006).

5.   S. Tarbouriech, G. Garcia, A.H. Glattfelder, *Advanced strategies in control systems with input and output constraints*, Vol. 346 of Lecture Notes in Control and Information Sciences. (Springer, Berlin Heidelberg New York London 2007).

6.   P. Hippe, *Windup in Control – Its Effects and Their Prevention.* (Springer, Berlin Heidelberg New York London 2006).

7.   L. Zaccarian, A.R. Teel, *Modern Anti-Windup Synthesis – Control Augmentation for Actuator saturation.* (Princeton University Press, Princeton, NJ and Oxford GB 2011).

8.   S. Tarbouriech, G. Garcia, J.M. Gomes, I. Queinnec, *Stability and Stabilization of Linear Systems with Saturating Actuators.* (Springer, London Dordrecht Heidelberg New York 2011).

9.   J.M. Biannic, S. Tarbouriech, Optimization and implementation of dynamic anti-windup compensators with multiple saturation in flight control systems. Control Eng. Pract. **17**(6), 703–713 (2009).

10. S. Galeani, S. Tarbouriech, M.C. Turner, L. Zaccarian, A tutorial on modern anti-windup design. Eur. J. Control **15**, 418–440 (2009).

11. J. Sofrony, *Anti-windup Compensation of Input Constrained Systems.* (VDM Verlag Dr. M, Saarbr 2009).

12. M.C. Turner, G. Herrman, I. Postlethwaite, *Anti-windup compensation and the control of input-constrained systems*, vol. 267 of Mathematical Methods for Robust and Nonlinear Control pp. 143–173 (2007).

13. A.R. Teel, N. Kapoor, *The L_2 anti-windup problem: its definition and solution.* In: Proceedings of the 4th European Control Conference, ECC'97, pp. 1897–1902. (Brussels, Belgium 1997).

14. G. Kaliora, A. Astolfi, Nonlinear control of feedforward systems with bounded signals. IEEE Trans. Autom. Control **49**, 1975–1990 (2004).

15. Z. Lin, T. Hu, Semi-global stabilization of linear systems subject to output saturation. Syst. Control Lett. **43**(3), 211–217 (2001).

16. Kreisselmeier, G.: Stabilization of linear systems in the presence of output measurement saturation. Systems & Control Letters **29**, 27–30 (1996)

17. J.K. Park, H.Y. Youn, Dynamic anti-windup based control method for state constrained systems. Automatica **39**, 1915–1922 (2003).

18. Y.Y. Cao, Z. Lin, B.M. Chen, An output feedback H_∞ controller for linear systems subject to sensor nonlinearities. IEEE Trans. Circuits Syst I **50**, 914–921 (2003).

19. Z. Zuo, J. Wang, L. Huang, Output feedback H_∞ controller design for linear discrete-time systems with sensor nonlinearities. IEE Proc. Control Theory Appl. **152**(1), 19–26 (2005).

20. R.B. Koplon, M.L.J. Hantus, E.D. Sontag, Observability of linear systems with saturated outputs. Linear Algebra Appl. **205**(1), 909–936 (1994).

21. Y. Xiao, Y.Y. Cao, Z. Lin, Robust filtering for discrete-time systems with saturation and its application to transmultiplexers. IEEE Trans. Signal Process. **52**(5), 1266–1277 (2004).

22. F. Yang, Y. Li, Set-membership filtering for systems with sensor saturation. Automatica **45**, 1896–1902 (2009).

23. J. Sofrony, M.C. Turner, *Coprime factor anti-windup for systems with sensor saturation.* In: Proceedings of the American Control Conference, pp. 3813–3818. (San Francisco, California 2011).

24. M.C. Turner, S. Tarbouriech, Anti-windup compensation for systems with sensor saturation: a study of architecture and structure. Int. J. Control **82**, 1253–1266 (2009).

25. M. Sassano, L. Zaccarian, Model recovery anti-windup for output saturated SISO linear closed loops. Syst. Control Lett. **85**, 109–117 (2015).

26. E. Chambon, L. Burlion, P. Apkarian, Time-response shaping using output to input saturation transformation. Int. J. Control **91**(3), 534–553 (2018).

Contents

Chapter 1
Prevention of Windup Caused by Saturating Sensors (Approach with Saturation Indicator)

1.1 Introductory Remarks

As demonstrated in the appendix, the reference behavior of a loop with input and/or output saturation can always be realized such that windup problems do not occur. This uses either model-based trajectory planning or the nonlinear model-based reference shaping filter described in Sect. A.1. Therefore, here and in all following chapters, attention is restricted to the problem of disturbance rejection.

When the sensor saturates, the loop is open and if the compensator contains integral action, the amplitude of this unstable element can grow beyond all limits causing the well-known windup effects. The windup prevention methods developed for input saturation are numerous and well documented and it therefore seems close at hand to apply these well-established solutions also to the case of saturating sensors. The first step with these well-known methods is a stabilization of the compensator during saturation (this prevents what is called *controller windup* in [1]), and if this does not remove all noxious effects of saturation (the possibly remaining effects are called *plant windup* in [1]), the linear part of the nonlinear loop is modified by a dynamic feedback of the difference between the saturated and the unsaturated signals (called the *Additional Dynamic Element* (ADE) in [1]) to guarantee stability of the nonlinear loop. Both approaches, however, are not applicable in the presence of output saturation.

Integral action can be viewed as a model process for constant or step-like changing disturbances. In the presence of persistent disturbances, the stabilization of the compensator during sensor saturation can lead to a malfunction of the compensator, i.e., it can prevent the disturbance rejection. If the reference signal is close to the sensor saturation limit, the input signal to the stabilized compensator can become so small that the amplitude of the generated control signal is not sufficient to get the output signal back into the linear range of the sensor. Therefore, **the integral action must stay active during output saturation!**

And the second step used with input saturation, namely, the modification of the linear part of the loop by a feedback of the difference between the limited and the unlimited signals, cannot be applied because this difference is unknown.

© The Author(s), under exclusive license to Springer Nature Switzerland AG 2021
P. Hippe, *Windup in Control Owing to Sensor Saturation*,
SpringerBriefs in Applied Sciences and Technology,
https://doi.org/10.1007/978-3-030-73133-5_1

The two above facts are probably one of the reasons why windup prevention in output-constrained systems is still not a well-researched field. Taking these facts into account one can design a new windup prevention scheme in the presence of output constraints.

1.2 Non-realizable Approach for SISO Systems

We consider strictly proper, asymptotically stable, completely controllable and observable minimum-phase LTI systems of the order n with manipulated input $u(t) \in \mathbb{R}^1$ and output $y(t) \in \mathbb{R}^1$. The transfer behavior of these linear systems is described by

$$y(s) = G(s)u(s) + G_d(s)d(s), \tag{1.1}$$

where $d(t) \in \mathbb{R}^1$ is a constant or step-like external disturbance. The transfer functions $G(s)$ and $G_d(s)$ have the forms

$$G(s) = \frac{N(s)}{D(s)}, \tag{1.2}$$

$$G_d(s) = \frac{N_d(s)}{D(s)}, \tag{1.3}$$

where $N(s)$ and $D(s)$ are coprime polynomials. The transfer behavior of the nominal linear controller is

$$u_C(s) = G_C(s)y_s(s), \tag{1.4}$$

where $y_s(t)$ is the output signal of the saturating sensor. The transfer function $G_C(s)$ has the form

$$G_C(s) = \frac{N_C(s)}{D_C(s)}. \tag{1.5}$$

For simplicity, it is assumed that both $D(s)$ and $D_C(s)$ are monic polynomials. The compensator is an *observer-based* controller which has been designed such that (i) it provides robust rejection of constant disturbances and (ii) with the feedback interconnection $u = -u_C$ and $y_s = y$, the closed-loop system is stable. This nominal compensator contains integral action so that one has

$$D_C(s) = sD_C^*(s). \tag{1.6}$$

It is assumed that the state observer is of minimum order so that the order of the compensator with integral action is $n_C = n$.

Fig. 1.1 Control loop with modified representation of the saturating nonlinearity

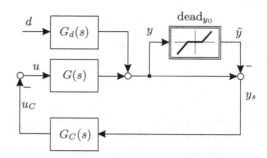

Remark 1.1 It is further assumed that apart from the integrator pole at the origin, the compensator is minimum phase also, i.e., $N_C(s)$ and $D_C^*(s)$ are Hurwitz polynomials. Given a minimum-phase system, such a compensator can always be designed [2].

The characteristic polynomial of the linear loop is

$$C_P(s) = N_C(s)N(s) + D_C(s)D(s). \tag{1.7}$$

It is convenient for this polynomial to be expressed as

$$C_P(s) = \Delta(s)\tilde{D}(s), \tag{1.8}$$

where $\Delta(s)$ is a Hurwitz polynomial of degree $n_C = n$ and $\tilde{D}(s)$ is a Hurwitz polynomial of degree n. In an observer-based compensator design, $\Delta(s)$ characterizes the dynamics of the state-plus-disturbance observer and $\tilde{D}(s)$ the dynamics of the system controlled by constant state feedback as, e.g., described in [1, 3].

The controller obtains feedback from a sensor, the output of which is related to the plant's output through the saturation nonlinearity $y_s = \text{sat}_{y_0}(y)$, where

$$\text{sat}_{y_0}(y) := \begin{cases} y_0 \text{ if } & y > y_0 > 0 \\ y \text{ if } & -y_0 \le y \le y_0 \\ -y_0 \text{ if } & y < -y_0. \end{cases} \tag{1.9}$$

Figure 1.1 shows the nonlinear loop with plant, compensator, and the saturating sensor. It has been drawn using the well-known identity

$$\text{sat}_{y_0}(y) = y - \text{dead}_{y_0}(y), \tag{1.10}$$

where

$$\tilde{y} = \text{dead}_{y_0}(y) := \begin{cases} y - y_0 \text{ if } & y > y_0 > 0 \\ 0 \text{ if } & -y_0 \le y \le y_0 \\ y + y_0 \text{ if } & y < -y_0. \end{cases} \tag{1.11}$$

The transfer function $G_L(s)$ of the linear part

$$y(s) = -G_L(s)\tilde{y}(s) \tag{1.12}$$

of the loop in Fig. 1.1 has the form

$$G_L(s) = -\frac{N_C(s)N(s)}{C_P(s)}. \tag{1.13}$$

Provided the disturbances are sufficiently small, nominal performance prevails. However, the output y can be driven beyond the range of the sensor for sufficiently large disturbance inputs. And if the transfer function (1.13) of the linear part of the nonlinear loop does not satisfy, e.g., the Circle Criterion (CC) stability of the nonlinear loop is no longer guaranteed.

Since the denominator of the compensator must remain unchanged (the integral action needs to stay active during saturation) a modification of the linear part of the loop can only be achieved by a re-shaping of the numerator polynomial of the compensator so that one has

$$G_{Cm}(s) = \frac{N_{Cm}(s)}{D_C(s)}. \tag{1.14}$$

This modified $N_{Cm}(s)$ then leads to a modified characteristic polynomial

$$C_{Pm}(s) = N_{Cm}(s)N(s) + D_C(s)D(s), \tag{1.15}$$

and to a modified

$$G_{Lm}(s) = -\frac{N_{Cm}(s)N(s)}{C_{Pm}(s)} \tag{1.16}$$

in (1.12). And if this modified transfer function satisfies the CC the nonlinear loop with sensor saturation behaves in a stable manner.

An appropriate modification of the linear part of the nonlinear loop can be achieved when feeding back the difference $y(s) - y_s(s)$ in an ADE as shown in Fig. 1.2. Indeed, when choosing the two Hurwitz polynomials $\tilde{\Omega}(s)$ and $\Omega(s)$ as

$$\tilde{\Omega}(s) = C_P(s) \tag{1.17}$$

and

$$\Omega(s) = C_{Pm}(s), \tag{1.18}$$

the transfer function in $y(s) = -G_{Lm}(s)\tilde{y}(s)$ in the loop of Fig. 1.2 has exactly the form (1.16). Thus, stability of the nonlinear loop is guaranteed. Unfortunately, the difference $y - y_s$ is not available.

In the next section, a means is presented to realize the stabilizing $G_{Lm}(s)$ in the constrained case.

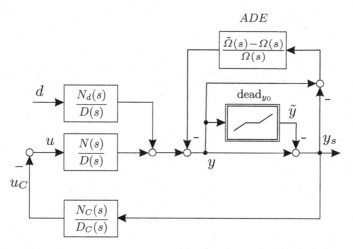

Fig. 1.2 Windup prevention with additional dynamic element

1.3 Realizable Approach for SISO Systems

The control system shown in Fig. 1.2 cannot be realized since the unconstrained output y is unknown. Figure 1.3 shows a realizable version of the windup prevention measure of Fig. 1.2. It uses a saturation indicator $\eta = \text{sind}_{y_0}(y_s)$ defined as

$$\text{sind}_{y_0}(y_s) := \begin{cases} 1 \text{ if } & y_s \geq y_0 > 0 \\ 0 \text{ if } & -y_0 < y < y_0 \\ 1 \text{ if } & y_s \leq -y_0. \end{cases} \tag{1.19}$$

Remark 1.2 If the sensor limit is not well defined for some reason, the saturation indicator might not function properly. In this case, one can place a model of the sensor saturation with a sufficiently smaller amplitude behind the sensor and replace y_s by the output of this model in the scheme of Fig. 1.3.

The system in Fig. 1.3 behaves as follows. As long as y is unlimited, the output of the saturation indicator is $\eta = 0$. However, when saturation is detected (which assumes accurate knowledge of y_0—see above remark), the output of the saturation indicator becomes unity. Thus, the saturation indicator detects the *presence*, but not the *extent*, of saturation. This is unlike the traditional trigger for anti-windup, dead(y), which detects the *presence and depth* of saturation.

Referring to Fig. 1.3 one can see that, when $|y| < y_0$, the nominal compensator $G_C(s)$ is active, but that the additional compensator $N_{Cd}(s)/D_C(s)$ is inactive (because the signal driving it is $\eta = 0$). Then, when $|y| > y_0$, the output of the saturation indicator is $\eta = 1$ which then activates the additional compensator. The numerator of the additional compensator transfer function is

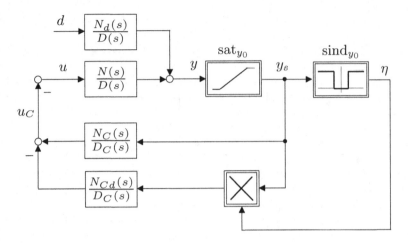

Fig. 1.3 Windup prevention with saturation indicator

$$N_{Cd}(s) = N_C(s) - N_{Cm}(s). \tag{1.20}$$

With this choice it follows that the overall controller transfer function (nominal controller plus anti-windup) is then $N_{Cm}(s)/D_C(s)$ as argued in the previous subsection. Thus, when saturation occurs, and assuming that $N_{Cm}(s)$ has been chosen correctly, the frequency response $G_{Lm}(j\omega)$ stays right of $(-1, j0)$.

Remark 1.3 In order to guarantee well-posedness of the system, and also to prevent high-frequency switching (chattering) in the system, the transfer function $N_{Cd}(s)/D_C(s)$ must be strictly proper.

The transfer function of the compensator must also satisfy the assumption of Remark 1.1. This can be achieved if the zeros of the system are contained in the characteristic polynomial (1.7), (1.8) of the loop. Then $G_{Lm}(j\omega)$ nearly always stays right of $(-1, j0)$ when assigning

$$N_{Cm}(s) = \Phi D(s), \tag{1.21}$$

with Φ the leading coefficient of $N_C(s)$.

This can be made plausible by the following arguments. Assume that $N^m(s)$ is a monic polynomial with the roots of $N(s)$ so that one has

$$N(s) = gN^m(s), \tag{1.22}$$

where g is the leading coefficient of $N(s)$. When the roots of $N(s)$ are part of $C_P(s)$, (1.7) shows that $D_C(s)$ will have the form

$$D_C(s) = D_{Cr}(s)N^m(s). \tag{1.23}$$

Then, with $N_{Cm}(s)$ according to (1.21) the transfer function (1.16) obtains the form

$$G_{Lm}(s) = -\frac{\Phi D(s)gN^m(s)}{\Phi D(s)gN^m(s) + D_{Cr}(s)N^m(s)D(s)} = -\frac{\Phi g}{\Phi g + D_{Cr}(s)}. \quad (1.24)$$

Since $D_{Cr}(s = 0) = 0$ the frequency response $G_{Lm}(j\omega)$ starts at -1 for $\omega = 0$. And if the magnitude of the Bode plot of $G_{Lm}(j\omega)$ has no overshoot, $G_{Lm}(j\omega)$ always stays right of the point $(-1, j0)$ in the complex plane. An overshoot in the magnitude of this Bode plot typically appears when the zeros of $C_P(s)$ are placed right of the zeros of $D(s)$. A good disturbance rejection, however, requires a left shift of the open-loop eigenvalues, so that $G_{Lm}(j\omega)$ has the desired property if (1.21) holds.

Remark 1.4 If one chose a design such that (1.21) did not lead to a $G_{Lm}(j\omega)$ staying right of $(-1, j0)$, $D(s)$ would have to be substituted by some appropriate monic Hurwitz polynomial $D_m(s)$ in Eq. (1.21). The search for a suitable $D_m(s)$, however, would then be a trial-and-error procedure.

Remark 1.5 The transients due to disturbance inputs d are, of course, not identical in the loops of Figs. 1.2 and 1.3, because in the constrained case the output y in Fig. 1.2 is smaller than the output y in Fig. 1.3. The stability behavior of both loops, however, appears to be identical. If, for a given nominal compensator and a polynomial $N_{Cm}(s)$ the loop in Fig. 1.2 does not exhibit an unstable behavior, there is no saturation limit y_0 for which the loop in Fig. 1.3 will become unstable and *vice versa*. And if for a given set $[N_C(s), D_C(s), N_{Cm}(s)]$ and a certain disturbance input there exists a saturation limit y_0 for which the loop in Fig. 1.2 exhibits limit cycling then there also exists a saturation limit for which the loop in Fig. 1.3 is limit cycling and *vice versa*.

The fact that both control loops in Figs. 1.2 and 1.3 have the same stability behavior is of further relevance. If the nominal compensator is designed for good disturbance rejection, i.e., the zeros of $C_P(s)$ are placed far left in the s-plane, the output signal u_C of the compensator can exceed the always-existing input signal limitations of the plant. Then the problem of windup prevention in the joint presence of input and output saturation has to be investigated. Such an anti-windup design cannot be carried out with the nonlinear elements in Fig. 1.3. Since the stability behavior of this configuration is the same as that of the loop in Fig. 1.2, the stability tests in the case of joint input and output constraints can be carried out when replacing the saturation indicator by an equivalent ADE (see Chap. 2).

Remark 1.6 It is obvious that in the MIMO case a choice of $N_{Cm}(s)$ according to (1.21) (where Φ would then be the highest row-degree-coefficient matrix of $N_C(s)$) will not lead to a transfer matrix $G_{Lm}(s)$ satisfying the CC. And even if, for a very special MIMO system, one succeeded in finding a matrix $N_{Cm}(s)$ such that $G_{Lm}(s)$ satisfied the CC there would not exist an equivalent MIMO version of the ADE. Therefore, the windup prevention method discussed in this chapter can only be applied to SISO systems.

The prevention of windup with the aid of the saturation indicator has two advantages. First, it follows the usual design paradigm of starting with an arbitrary (though minimum phase in the sense of Remark 1.1) compensator and a subsequent addition of measures to avoid windup caused by the limited sensor. Second, though a rigorous proof of stability is not known to date (and there might never be one), the scheme in Fig. 1.3 works perfectly. Even if the eigenvalues of the plant are shifted far into the left half plane by the nominal compensator the behavior of the nonlinear loop is always stable. This will also be demonstrated in Example 1.2.

The presented solution lacks a rigorous proof of stability (what can be tolerated because it works perfectly). Its major drawback, however, is its inapplicability to MIMO systems. Therefore, later we will consider a much simpler approach for which a strict proof of stability exists even in the MIMO case.

1.4 Résumé of the Design Steps

We start with a strictly proper minimum-phase SISO system of the order n with transfer function

$$G(s) = \frac{N(s)}{D(s)}, \tag{1.25}$$

for which an observer-based compensator with integral action of the order $n_C = n$ is designed. Its transfer behavior is

$$G_C(s) = \frac{N_C(s)}{D_C(s)}, \tag{1.26}$$

with $D_C(s) = s D_C^*(s)$. The compensator assures a desired amount of disturbance rejection while both polynomials $N_C(s)$ and $D_C^*(s)$ are Hurwitz. This can be achieved by making the zeros of $N(s)$ part of the design polynomial $C_P(s) = \Delta(s)\tilde{D}(s)$ and by shifting the remaining zeros of $C_P(s)$ into the left half s-plane. If

$$G_L(s) = -\frac{N_C(s)N(s)}{C_P(s)} \tag{1.27}$$

does not satisfy the CC, stability of the nonlinear loop in Fig. 1.1 is not guaranteed. A stable behavior is obtained when realizing the loop in the form of Fig. 1.3. The polynomial

$$N_{Cd}(s) = N_C(s) - N_{Cm}(s) \tag{1.28}$$

is obtained from

$$N_{Cm}(s) = \Phi D(s), \tag{1.29}$$

where Φ is the leading coefficient of $N_C(s)$.

Example 1.1 First, let us consider a very simple system with transfer functions

$$G(s) = \frac{s+4}{s^2 + 2s + 2},$$

and

$$G_d(s) = \frac{8s + 20}{s^2 + 2s + 2}.$$

The disturbance input is $d(t) = 1(t)$ and the sensor saturation limit is $y_0 = 0.4$. The design parameters of the compensator are chosen as

$$\Delta(s) = (s+5)(s+4),$$

to compensate the plant zero at $s = -4$ and as

$$\tilde{D}(s) = (s+5)^2.$$

The transfer function of the compensator with integral action leading to a $C_P(s)$ with the above-defined zeros is

$$G_C(s) = \frac{13s^2 + 73s + 125}{s^2 + 4s}.$$

Since the open-loop gain $y(s) = -G_L(s)\tilde{y}(s)$ in the loop of Fig. 1.1 does not satisfy the CC one needs measures for the prevention of windup. With $N_{Cm}(s) = 13D(s)$ the transfer function

$$G_{Lm}(s) = -\frac{N_{Cm}(s)N(s)}{N_{Cm}(s)N(s) + D_C(s)D(s)}$$

satisfies the CC and with

$$\frac{N_{Cd}(s)}{D_C(s)} = \frac{47s + 99}{s^2 + 4s},$$

the loop in Fig. 1.3 has the disturbance behavior shown in Fig. 1.4. The broken line represents the real output y of the system and the full line the output y_s of the sensor.

The second system is of higher order and the disturbance rejecting compensators are real high-gain examples.

Example 1.2 Consider a plant described by its transfer functions

$$G(s) = \frac{6.25s^2 + 37.5s + 50}{s^4 + 8s^3 + 39s^2 + 62s + 50}$$

and

Fig. 1.4 Disturbance
transients of Example 1.1

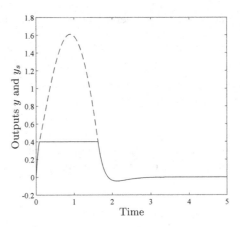

$$G_d(s) = \frac{20s^3 + 80s^2 + 170s + 300}{s^4 + 8s^3 + 39s^2 + 62s + 50}.$$

There is a step-like input disturbance $d(t) = 1(t)$ and the sensor saturates at $y_0 = 0.2$. The compensator with integral action is designed to obtain a very good disturbance rejection. Its design parameters are

$$\tilde{D}(s) = (s + 20)^4,$$

and since the zeros of the plant are located at $s = -2$ and $s = -4$, the observer polynomial is chosen as

$$\Delta(s) = (s + 20)^2(s + 2)(s + 4).$$

This leads to the transfer function

$$G_C(s) = \frac{810.4s^4 + 24891.2s^3 + 382880.96s^2 + 3071104s + 10240000}{(s^2 + 112s)(s + 2)(s + 4)}$$

of the compensator. With this compensator the nonlinear loop is limit cycling for the existing output saturation. Choosing

$$N_{Cm}(s) = 810.4D(s),$$

the transfer function (1.16) satisfies the CC and the numerator (1.20) has the form

$$N_{Cd}(s) = 18408s^3 + 351275.36s^2 + 3020859.2s + 10199480.$$

Fig. 1.5 Disturbance transients with sensor saturation

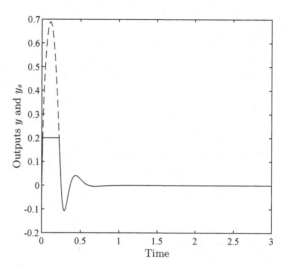

Realizing the loop in the form of Fig. 1.3 the disturbance transients shown in Fig. 1.5 result. The broken line shows the plant output y and the full line the output y_s of the sensor.

If one had chosen

$$N_{Cm}(s) = 810.4(s^4 + 8s^3 + 67s^2 + 236s + 480),$$

the resulting frequency response $G_{Lm}(s)$ would indicate the danger of limit cycling in the loop. Indeed, the loop realized according to Fig. 1.2 starts limit cycling for all output saturation limits $y_0 < 0.0018$ and the loop realized according to Fig. 1.3 starts limit cycling for all $y_0 < 0.018$ which demonstrates the claims of Remark 1.5.

If a disturbance drives the output beyond the saturation limit of the sensor, the amplitude of the actual system output can attain large amplitudes (much bigger than they would develop in the unconstrained case) and it also takes much longer before the disturbance is compensated. Therefore, one could try to obtain an even better disturbance attenuation such that the disturbance does not cause the sensor to saturate.

Here, this is accomplished when selecting the design parameters as

$$\tilde{D}(s) = (s + 34)^4$$

and

$$\Delta(s) = (s + 34)^2 (s + 2)(s + 4).$$

This leads to the transfer function

Fig. 1.6 Disturbance
transients with faster control

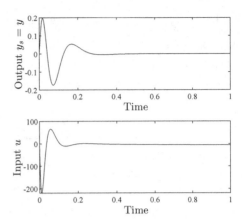

$$G_C(s) =$$
$$\frac{2517.28s^4 + 124539.84s^3 + 3205254.08s^2 + 43616439.04s + 247168706.56}{(s^2 + 196s)(s + 2)(s + 4)}$$

of the compensator. With

$$N_{Cm}(s) = 2517.28D(s),$$

the transfer function $G_{Lm}(s)$ again satisfies the CC. The polynomial (1.20) now has
the form

$$N_{Cd}(s) = 104401.6s^3 + 3107080.16s^2 + 43460367.68s + 247042842.56.$$

Figure 1.6 shows the output $y_s = y$ of the system and the input u generated by this
compensator. It is obvious that the amplitude of the input signal exceeds the always-
existing input limitations. When an input saturation with $u_0 = 30$ is inserted in the
loop of Fig. 1.3 it starts limit cycling in simulations. In the loop of Fig. 1.2, limit
cycling starts for input saturation limits $u_0 \leq 27$.

Since the amplitude of the attacking disturbances is usually not precisely known,
there will be applications where it cannot be ruled out that also input saturation
becomes active. Then, windup prevention in the joint presence of input and output
saturation is a necessity.

1.5   Discussion of the Results

Although the stability of the loop in Fig. 1.3 cannot be proven rigorously the author is 100% sure that the control structure with saturation indicator is stable. All examples ever examined have shown this, and these examples have been very extreme in many cases, as in Example 1.2.

If the fast compensator of Example 1.2 had been used in a loop with input saturation, one would have started with a stabilization of the compensator during saturation by using, e.g., the observer technique (i.e., $D_C(s)$ is substituted by $A(s)$ during saturation). This prevents the so-called controller windup, but it would not have stabilized the nonlinear loop. Due to the extreme left shift of the eigenvalues by state feedback (characterized by the zeros of the polynomial $\tilde{D}(s)$) an intensive plant windup would have existed in addition. This would have led to limit cycling for all input saturation limits $u_0 < 69$. This plant windup could, of course, have been prevented with the aid of a suitably designed ADE [1].

Comparing this two-step procedure for windup prevention with the method used in this chapter it is hard to believe that the simple scheme of Fig. 1.3 stabilizes the nonlinear loop, especially with integral action active during saturation. However, it does!

Stability of the scheme in Fig. 1.3 is achieved by a polynomial $N_{Cm}(s)$ assuring that $G_{Lm}(s)$ in (1.16) satisfies the CC. However, if no additional properties existed, skepticism concerning the stability behavior of the scheme would be justified.

To discuss the importance of the first property it is convenient to inspect the nonlinear loop in Fig. 1.7. It is a non-realizable version of the loop in Fig. 1.3 and exhibits an identical behavior for $V = y_0$.

This first property concerns the transfer function $\frac{N_{Cd}(s)}{D_C(s)}$. If it is not strictly proper, a switching by the sign element in Fig. 1.7 could cause an immediate backlash of the

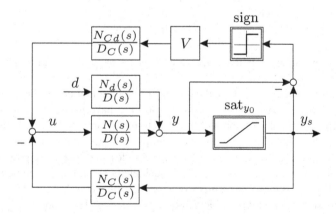

Fig. 1.7 Equivalent scheme to Fig. 1.3 for $V = y_0$

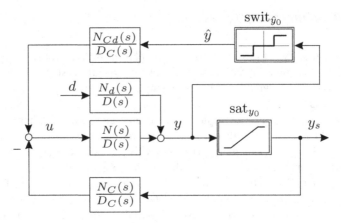

Fig. 1.8 Another equivalent scheme to Fig. 1.3 for $\hat{y}_0 = y_0$

input signal u which is then counteracted by the linear compensator. This can lead to high-frequency oscillation or chattering.

Indeed, if the polynomial $N_{Cm}(s)$ in Example 1.1 is modified to $N_{Cm}(s) = 31D(s)$ the transfer function $G_{Lm}(s)$ still satisfies the CC, but the simulation stops because of high-frequency oscillations. In Example 1.2, the factor Φ in (1.21) can be increased without such effects. As long as $G_{Lm}(s)$ satisfies the CC no chattering occurs. The reason for this seems to be the low-pass property of the system with its difference degree $\kappa = 2$.

The second, even more important, property of the scheme in Fig. 1.7 is the gain $V = y_0$. The motivation for investigating the role of V becomes clear when inspecting the block diagram in Fig. 1.8. It incorporates a switching nonlinearity $\hat{y} = \text{swit}_{\hat{y}_0}(y)$ where

$$\text{swit}_{\hat{y}_0}(y) := \begin{cases} \hat{y}_0 & \text{if} \quad y > \hat{y}_0 > 0 \\ 0 & \text{if} \; -\hat{y}_0 \le y \le \hat{y}_0 \\ -\hat{y}_0 & \text{if} \quad y < -\hat{y}_0. \end{cases} \tag{1.30}$$

For $\hat{y}_0 = y_0$, this scheme exhibits the same behavior as the loop in Fig. 1.3 and the loop in Fig. 1.7 for $V = y_0$.

The CC allows statements about the stability of loops containing isolated nonlinear elements bounded by two straight lines. These lines have the slopes one and zero for the saturation, the dead zone and the switching element in Fig. 1.8. If the CC is satisfied, stability is guaranteed for arbitrary amplitudes y_0 of the saturation, arbitrary breadths y_0 of the dead zone, and arbitrary amplitudes $\hat{y}_0$ of the switching element.

Here in the loop of Fig. 1.8 we have two such elements. If one developed a proof for the stability of this loop, the results would be extremely conservative, because they would assure stability for all possible amplitudes y_0 and $\hat{y}_0$. The loops of Figs. 1.3 and 1.8, however, are only equivalent when y_0 and $\hat{y}_0$ coincide. And indeed, for $\hat{y}_0 \ne y_0$ stability is no longer guaranteed.

In Example 1.1, the compensator has a modest gain. When the sensor saturates at $y = 0.1$ the loop in Fig. 1.8 becomes unstable in simulations with $\hat{y}_0 = 0.127$, i.e., the amplitudes of the two nonlinear elements can only vary for about 26% before the disturbance transients no longer decay, but increase. In Example 1.2, this is even more pronounced. Due to the high-gain compensator, smaller differences between y_0 and $\hat{y}_0$ already cause an unstable behavior. If the sensor saturates at $y_0 = 0.2$, the loop with the "slower" of the two compensators leads to unstable disturbance transients either for $\hat{y}_0 = 0.2127$ or for $\hat{y}_0 = 0.169$.

This demonstrates that standard methods are not suitable to prove the stability of the anti-windup control with saturation indicator. One would need special tests also taking into account the amplitudes of the nonlinearities.

References

1. P. Hippe, *Windup in Control—Its Effects and Their Prevention* (Springer, Berlin, Heidelberg, New York, London, 2006)
2. G.C. Goodwin, Private communication (2011)
3. P. Hippe, J. Deutscher, *Design of Observer-based Compensators—From the Time to the Frequency Domain* (Springer, Berlin, Heidelberg, New York, London, 2009)

Chapter 2
Prevention of Windup Caused by Saturating Sensors and Actuators (With Saturation Indicator)

2.1 Introductory Remarks

The method introduced in Chap. 1 allows to use an arbitrary controller for the rejection of step-like or constant disturbances and if this controller causes stability problems in case of sensor saturation, its numerator is modified with the aid of a Saturation Indicator (SI) such that the linear part of the loop satisfies the Circle Criterion (CC).

This allows a nearly perfect attenuation of persistent disturbances in the presence of saturating sensors. However, better disturbance rejection has to be bought by increased plant input signals so that the always-existing input limitations can also be violated. This is why it can be necessary to assure that joint input and output saturation do not destabilize the loop.

The SI is a switch located behind the sensor, and therefore it cannot be used in the stability analysis of a loop with input *and* output saturation. However, it has been demonstrated in Chap. 1 that the stability behavior of the SI is the same as that of an appropriately designed Additional Dynamic Element (ADE). Thus, in the stability analysis, the SI can be substituted by the equivalent ADE. And if the anti-windup scheme with this (not realizable) ADE is shown to work in a stable manner, the realizable anti-windup scheme with SI is also stable.

Whereas in the presence of mere output constraints the compensator was allowed to be arbitrarily fast, this is no longer the case if input and output saturate at the same time. Therefore, the dynamics of the nominal compensator have to be reduced in the following investigations with the consequence of a reduced disturbance attenuation. In view of this, one could circumvent the effort of assuring stability in the presence of joint input and output limitations by reducing the dynamics of the nominal loop right from the beginning to an extent that the expected disturbances do not cause input saturation at all. Disturbances, however, have the unpleasant property that neither their time of attack nor their actual amplitude are predictable.

Therefore, the following design effort may sometimes be necessary.

2.2   Problem Formulation

We consider asymptotically stable, strictly proper, minimum-phase LTI systems of the order n with manipulated input $u_s(t) \in \mathbb{R}^1$ and output $y(t) \in \mathbb{R}^1$. The transfer behavior of these systems is described by

$$y(s) = G(s)u_s(s) + G_d(s)d(s), \qquad (2.1)$$

where $d(t) \in \mathbb{R}^1$ is a constant or step-like disturbance. The transfer functions $G(s)$ and $G_d(s)$ are

$$G(s) = \frac{N(s)}{D(s)}, \qquad (2.2)$$

$$G_d(s) = \frac{N_d(s)}{D(s)}, \qquad (2.3)$$

where $N(s)$ and $D(s)$ are coprime polynomials. The linear compensator is described by

$$u_C(s) = G_C(s)y_s(s), \qquad (2.4)$$

where y_s is the output of the sensor. The transfer function $G_C(s)$ is represented as

$$G_C(s) = \frac{N_C(s)}{D_C(s)}. \qquad (2.5)$$

It is assumed that the compensator is an observer-based controller which has been designed such that (i) it provides robust rejection of constant disturbances and (ii) with the feedback interconnection $u_s = u = -u_C$ and $y_s = y$, the closed-loop system is stable. For simplicity, we further assume that $D(s)$ and $D_C(s)$ are monic polynomials. The nominal compensator contains integral action so that one has

$$D_C(s) = sD_C^*(s). \qquad (2.6)$$

It is assumed that the state observer is of minimum order so that the order of the compensator with integral action is $n_C = n$. It is also assumed that apart from the integrator pole at the origin, the compensator is minimum phase, i.e., $N_C(s)$ and $D_C^*(s)$ are Hurwitz polynomials.

The characteristic polynomial of the linear loop is

$$C_P(s) = N_C(s)N(s) + D_C(s)D(s). \qquad (2.7)$$

In observer-based control, this has the form

$$C_P(s) = \Delta(s)\tilde{D}(s), \qquad (2.8)$$

where $\Delta(s)$ is a Hurwitz polynomial of degree n_C (characterizing the dynamics of the state-plus-disturbance observer) and $\tilde{D}(s)$ is a Hurwitz polynomial of degree n (characterizing the dynamics assigned by state feedback) [1, 2].

The measured output y_s is a saturated version of y. The saturation nonlinearity $y_s = \mathrm{sat}_{y_0}(y)$ is defined by

$$\mathrm{sat}_{y_0}(y) := \begin{cases} y_0 & \text{if } \quad y > y_0 > 0 \\ y & \text{if } -y_0 \leq y \leq y_0 \\ -y_0 & \text{if } \quad y < -y_0. \end{cases} \qquad (2.9)$$

The input u_s is a saturated version of u. The nonlinearity $u_s = \mathrm{sat}_{u_0}(u)$ is defined by

$$\mathrm{sat}_{u_0}(u) := \begin{cases} u_0 & \text{if } \quad u > u_0 > 0 \\ u & \text{if } -u_0 \leq u \leq u_0 \\ -u_0 & \text{if } \quad u < -u_0. \end{cases} \qquad (2.10)$$

If saturation takes place at the input or the output, the loop is open and the integral action in the compensator can wind up. First, it must be assured that neither input nor output saturation alone destabilizes the loop.

2.2.1 Output Saturation

This problem was solved in Chap. 1 so that we can use the results obtained there. The (realizable) scheme for windup prevention in case of sensor saturation has the form of Fig. 2.1 (see also Fig. 1.3). The numerator polynomial

$$N_d(s) = N_C(s) - N_{Cm}(s) \qquad (2.11)$$

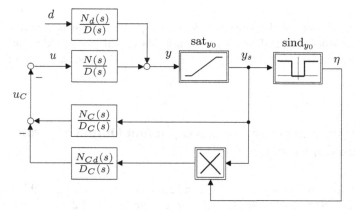

Fig. 2.1 Windup prevention with saturation indicator

is obtained from

$$N_{Cm}(s) = \Phi D(s), \tag{2.12}$$

where Φ is the leading coefficient of $N_C(s)$. When the compensator is designed such that the zeros of the system are compensated, (2.12) assures a linear part of the loop satisfying the CC (see also the corresponding discussion in Chap. 1). The loop in Fig. 2.1 also assures Linear Performance Recovery (LPR).

2.2.2 Input Saturation

When only the input signal is saturated one can use the well-established methods for windup prevention. The most straightforward way to prevent integral windup is the *observer technique* [1]. During input saturation it replaces the non-Hurwitz denominator polynomial $D_C(s)$ by the Hurwitz polynomial $\Delta(s)$. Using

$$N_u(s) = D_C(s) - \Delta(s) \tag{2.13}$$

for a realization

$$u_C(s) = \frac{N_u(s)}{\Delta(s)} u_s(s) + \frac{N_C(s)}{\Delta(s)} y(s) \tag{2.14}$$

of the compensator, controller windup due to input saturation does no longer occur and this also assures LPR. If the resulting linear part

$$u(s) = -G_{Lu}(s)u_s(s) \tag{2.15}$$

of the loop with

$$G_{Lu}(s) = \frac{\tilde{D}(s) - D(s)}{D(s)} \tag{2.16}$$

violates the CC there is the danger of *plant windup* [1]. This can be prevented by an ADE. If both the sensor and the actuator saturate the closed-loop dynamics achievable are quite restricted (see Sect. 2.3) so that the danger of plant windup is no problem here.

2.3 Windup Prevention when Input and Output Saturate Simultaneously

The conditions presented in Sects. 2.2.1 and 2.2.2 assure that windup problems do not occur when either the input u or the output y saturates. Figure 2.2 shows the nonlinear loop with the above-discussed measures for the prevention of windup due to input or

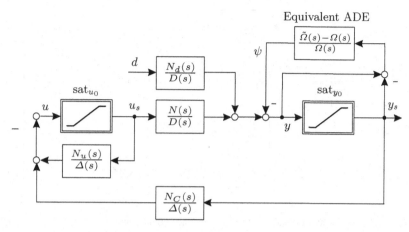

Fig. 2.2 Windup prevention if either the input or the output saturate

output saturation. In this block diagram, the SI is substituted by the equivalent ADE (see Sect. 1.2) since stability investigations of a nonlinear loop with input and output constraints are not feasible in the presence of the SI.

To assure a stable behavior when both saturation elements become active, the linear part of the loop in Fig. 2.2 must satisfy the CC. The system has one input and one output. In a representation of the loop of Fig. 2.1 in standard form (linear part and isolated nonlinearity), the linear part

$$\begin{bmatrix} u(s) \\ y(s) \end{bmatrix} = -G_{Luy}(s) \begin{bmatrix} u_s(s) \\ y_s(s) \end{bmatrix} \tag{2.17}$$

is a system with two inputs and two outputs with

$$G_{Luy}(s) = \begin{bmatrix} \dfrac{N_u(s)}{\Delta(s)} & \dfrac{N_C(s)}{\Delta(s)} \\ -\dfrac{N(s)\Omega(s)}{D(s)\tilde{\Omega}(s)} & -\dfrac{\tilde{\Omega}(s) - \Omega(s)}{\tilde{\Omega}(s)} \end{bmatrix}. \tag{2.18}$$

It is obvious that a compensator designed for very good disturbance attenuation will not lead to a transfer matrix (2.18) satisfying the CC. Instead, the allowable left shift of the eigenvalues is now restricted. This is not surprising because the control task is not easy. The system with step-like disturbance input and the compensator with integral action are in an open-loop situation at both sides for an unknown period of time before they are finally connected again. And this must happen with guaranteed stability. Consequently, the compensator should not react too nervously.

An ADE provides additional degrees of freedom for improving the control scheme of Fig. 2.2. At the output side an ADE cannot be realized since the difference $y - y_s$ is unknown (the one in Fig. 2.2 mimics the stability characteristic of the SI). At the

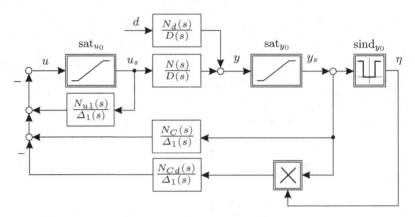

Fig. 2.3 Windup prevention when both the input and the output saturate

input side an ADE can be applied. It is also parametrized by two Hurwitz polynomials $\tilde{\Omega}_u(s)$ and $\Omega_u(s)$ having the same degree and the same leading coefficient. For $\Omega_u(s) = \Delta(s)$, the ADE at the input side can be realized in the compensator (see [1], Sect. 4.2). When denoting $\tilde{\Omega}_u(s) = \Delta_1(s)$ the loop of Fig. 2.2 with an ADE at the input side obtains the form of Fig. 2.3 (now with SI in place). The polynomial $\Delta_1(s)$ is a monic Hurwitz polynomial of the same degree as $\Delta(s)$.

For the following discussions of stability, the block diagram of Fig. 2.2 is relevant, however, with $\Delta(s)$ substituted by $\Delta_1(s)$ and $N_u(s)$ substituted by

$$N_{u1}(s) = D_C(s) - \Delta_1(s). \tag{2.19}$$

After the introduction of the input ADE the transfer matrix of the linear part in the block diagram of Fig. 2.2 has the form

$$G_{Luy}^1(s) = \begin{bmatrix} \dfrac{N_{u1}(s)}{\Delta_1(s)} & \dfrac{N_C(s)}{\Delta_1(s)} \\ -\dfrac{N(s)\Omega(s)}{D(s)\tilde{\Omega}(s)} & -\dfrac{\tilde{\Omega}(s) - \Omega(s)}{\tilde{\Omega}(s)} \end{bmatrix}. \tag{2.20}$$

For the test of the MIMO CC, it is convenient to use the time-domain representation of (2.17). Given the time-domain representations of the linear systems in Fig. 2.2 (with $\Delta(s) = \Delta_1(s)$ and (2.19)), namely,

$$\begin{aligned} \dot{x}(t) &= A_p x(t) + B_p u_s(t) \\ y(t) &= C_p x(t) - \psi(t), \end{aligned} \tag{2.21}$$

$$\begin{aligned} \dot{\xi}(t) &= A_\xi \xi(t) + B_\xi (y(t) - y_s(t)) \\ \psi(t) &= C_\xi \xi(t), \end{aligned} \tag{2.22}$$

and

$$\dot{z}(t) = A_c z(t) + \begin{bmatrix} B_{cu} & B_{cy} \end{bmatrix} \begin{bmatrix} u_s(t) \\ y_s(t) \end{bmatrix}$$

$$u(t) = -C_c z(t) - \begin{bmatrix} 0 & D_{cy} \end{bmatrix} \begin{bmatrix} u_s(t) \\ y_s(t) \end{bmatrix},$$

(2.23)

the time-domain representation of (2.17) with $G_{Luy}(s) = G^1_{Luy}(s)$ has the form

$$\begin{bmatrix} \dot{x}(t) \\ \dot{\xi}(t) \\ \dot{z}(t) \end{bmatrix} = \begin{bmatrix} A_p & 0 & 0 \\ B_\xi C_p & A_\xi - B_\xi C_\xi & 0 \\ 0 & 0 & A_c \end{bmatrix} \begin{bmatrix} x(t) \\ \xi(t) \\ z(t) \end{bmatrix} + \begin{bmatrix} B_p & 0 \\ 0 & -B_\xi \\ B_{cu} & B_{cy} \end{bmatrix} \begin{bmatrix} u_s(t) \\ y_s(t) \end{bmatrix}$$

$$\begin{bmatrix} u(t) \\ y(t) \end{bmatrix} = \begin{bmatrix} 0 & 0 & -C_c \\ C_p & -C_\xi & 0 \end{bmatrix} \begin{bmatrix} x(t) \\ \xi(t) \\ z(t) \end{bmatrix} + \begin{bmatrix} 0 & -D_{cy} \\ 0 & 0 \end{bmatrix} \begin{bmatrix} u_s(t) \\ y_s(t) \end{bmatrix}.$$

(2.24)

A high-gain compensator will not lead to a $G^1_{Luy}(s)$ satisfying the CC. Instead, one must look for suitable zeros of the polynomials $\tilde{D}(s)$ and $\Delta(s)$ so that $G^1_{Luy}(s)$ satisfies the CC.

The design of the scheme in Fig. 2.3 is not straightforward. It requires an iteration process where the starting value is a compensator such that $G_{Luy}(s)$ satisfies the CC. When shifting the zeros of $\tilde{D}(s)$ and $\Delta(s)$ further left $G_{Luy}(s)$, which coincides with $G^1_{Luy}(s)$ for $\Delta_1(s) = \Delta(s)$, will no longer satisfy the CC. This can (probably) be fixed by an appropriate polynomial $\Delta_1(s)$. This requires an "optimization" to find the best $\Delta_1(s)$. If the "optimization" was successful, the eigenvalues of the linear loop can be shifted further left, if not, they have to be shifted right until a polynomial $\Delta_1(s)$ can be obtained for which $G^1_{Luy}(s)$ satisfies the CC. This is a trial-and-error process which is complicated by the basic problems of a pole-placing design. It leads, however, to the best solution achievable.

Remark 2.1 For SISO systems with transfer function $G(s)$, the CC can be tested by simply plotting the frequency response $G(j\omega)$. If this frequency response stays right of -1 in the complex plane the CC is satisfied. Although we are investigating SISO systems, here $G_{Luy}(s)$ and $G^1_{Luy}(s)$ are 2×2 matrices and when investigating MIMO systems with p inputs and outputs later, the corresponding transfer matrices will have the dimensions $2p \times 2p$. The author does not know any standard software allowing to check whether a $p \times p$ transfer matrix with $p > 1$ satisfies the CC or not. Fortunately, a former colleague of his developed an m-file to test the CC for MIMO systems [3]. This m-file also generates an appropriate multiplier so that the results are less conservative. Since the outcome of this m-file is simply "yes" or "no," an "optimization" to obtain improved results is a tedious trial-and-error process. It can be solved by starting from an upper sector limit < 1 assured for the starting values and then using a Monte Carlo method to gradually widen the sector up to the desired value 1. A perfect solution would be a systematic design of the compensator such that $G^1_{Luy}(s)$ satisfies the CC. Such methods, however, are based on a feedback of the complete state of the system. The dynamics of $G^1_{Luy}(s)$ are of the order $4n$ while the compensator is only of the order n. This horrendous increase in order is due to the substitution of the SI by an equivalent ADE. With the methods discussed in Chaps. 4 and 5 this increase in order is reduced to $2n$ and also the pole-placing design is dramatically simplified. The necessity of using a Monte Carlo method, however, remains.

2.4 Résumé of the Design Steps

We start with a system and a compensator as described in Sect. 1.4. The nominal compensator is characterized by its polynomials $N_C(s)$ and $D_C(s)$ leading to the characteristic polynomial

$$C_P(s) = N_C(s)N(s) + D_C(s)D(s) = \Delta(s)\tilde{D}(s) \tag{2.25}$$

of the linear closed loop. The modified compensator has the same denominator but a modified numerator polynomial $N_{Cm}(s)$ (see (2.12)). This leads to the modified characteristic polynomial

$$C_{Pm}(s) = N_{Cm}(s)N(s) + D_C(s)D(s). \tag{2.26}$$

These two characteristic polynomials define an ADE in Fig. 2.2 with

$$\tilde{\Omega}(s) = C_P(s) \quad \text{and} \quad \Omega(s) = C_{Pm}(s), \tag{2.27}$$

such that for $\Delta_1(s) = \Delta(s)$ the stability behaviors of the loops in Figs. 2.2 and 2.3 are the same (see the discussions in Sects. 1.2 and 1.3).

Look for polynomials $\Delta(s)$ and $\tilde{D}(s)$ such that the transfer matrix $G_{Luy}(s)$ in (2.18) satisfies the CC. As demonstrated in Chap. 4 this is not always possible. Then shift the zeros of $\Delta(s)$ and $\tilde{D}(s)$ further left (i.e., increase the speed of the nominal control) and try to obtain a polynomial $\Delta_1(s)$ such that also $G^1_{Luy}(s)$ (see (2.20)) satisfies the CC. If this is easily possible increase the left shift and try anew. If there is no $\Delta_1(s)$, decrease the left shift until the fastest nominal control is obtained (see also Remark 2.1).

Example 2.1 As a demonstrating example consider the system of Example 1.2 with its transfer functions

$$G(s) = \frac{N(s)}{D(s)} = \frac{6.25s^2 + 37.5s + 50}{s^4 + 8s^3 + 39s^2 + 62s + 50},$$

and

$$G_d(s) = \frac{N_d(s)}{D(s)} = \frac{20s^3 + 80s^2 + 170s + 300}{s^4 + 8s^3 + 39s^2 + 62s + 50}.$$

In addition to an output saturation with $y_0 = 0.5$ there is also an input saturation with $u_0 = 5$. The compensators designed in Example 1.2 are far from satisfying either $G_{Luy}(s)$ or $G^1_{Luy}(s)$. A choice of $\tilde{D}(s) = (s+5)^2(s+2)^2$ and $\Delta(s) = (s+2)(s+4)(s+2)^2$ leads to a compensator with transfer function

$$G_C(s) = \frac{1.6s^4 + 3.2s^3 + 42.56s^2 + 73.6s + 64}{s^4 + 16s^3 + 68s^2 + 80s}, \tag{2.28}$$

and consequently to the polynomial $N_{Cm}(s) = 1.6D(s)$. With this compensator the transfer matrix $G_{Luy}(s)$ satisfies the CC, i.e., the loop of Fig. 2.3 has a stable behavior for $\Delta_1(s) = \Delta(s)$.

When performing the above-described iterative design, the polynomial

$$\Delta_1(s) = s^4 + 6.89743s^3 + 21.40365s^2 + 33.3590s + 18.1310 \tag{2.29}$$

can be obtained. It allows a left shift of the zeros of $\tilde{D}(s)$ and $\Delta(s)$ to obtain $\tilde{D}(s) = (s+5)^2(s+3)^2$ and $\Delta(s) = (s+2)(s+4)(s+3)^2$. This leads to a nominal compensator with transfer function

$$G_C(s) = \frac{7.68s^4 + 54.4s^3 + 254.88s^2 + 449.6s + 324}{s^4 + 20s^3 + 92s^2 + 112s}, \tag{2.30}$$

so that $N_{Cm}(s)$ has the form $N_{Cm}(s) = 7.68D(s)$. With this compensator and the above $\Delta_1(s)$ the transfer matrix $G^1_{Luy}(s)$ satisfies the CC.

It is obvious that compensator (2.30) allows a better disturbance attenuation than compensator (2.28). Figure 2.4 shows in broken lines the disturbance step response of the linear loop with the compensator (2.28) and in full lines the one with the compensator (2.30). Though the improvement is not spectacular, the peak value of the disturbance transient can be reduced by 27%.

Fig. 2.4 Disturbance step responses of the linear loop

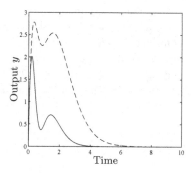

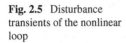

Fig. 2.5 Disturbance transients of the nonlinear loop

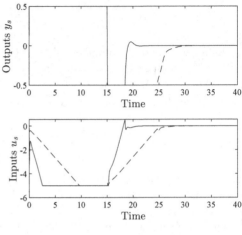

Fig. 2.6 Disturbance transients with SI and ADE

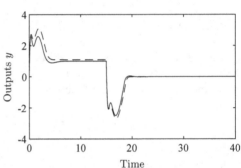

As the signal $u(t)$ does not exhibit a distinct overshoot behavior for step-like inputs, the transients of the nonlinear loop are demonstrated for a disturbance input $d(t) = 1(t) - 1(t - 15)$ which causes a saturation both at the input and the output during the first 15 s. The broken lines in Fig. 2.5 show the reactions of the nonlinear loop in Fig. 2.3 (with compensator (2.28) and $\Delta_1(s) = \Delta(s)$) to this disturbance input and the full lines depict the reactions with compensator (2.30) and $\Delta_1(s)$ as in (2.29) to the same input.

The tests of the CC were based on the non-realizable structure shown in Fig. 2.2 but the simulations use the SI according to Fig. 2.3, of course. Figure 2.6 shows the outputs y of the nonlinear loop with ADE (broken lines) and with SI (full lines) for the compensator (2.30) with (2.29). This is an indication that the replacement of the SI by the ADE in the optimization makes sense.

References

1. P. Hippe, *Windup in Control—Its Effects and Their Prevention* (Springer, Berlin, Heidelberg, New York, London, 2006)
2. P. Hippe, J. Deutscher, *Design of Observer-based Compensators—From the Time to the Frequency Domain* (Springer, Berlin, Heidelberg, New York, London, 2009)
3. G. Schmitt, Frequency domain evaluation of circle criterion, Popov criterion and off-axis circle criterion in the MIMO case. Int. J. Control **72**, 1299–1309 (1999)

Chapter 3
Prevention of Windup Caused by Saturating Sensors (Compensating Approach)

3.1 Introductory Remarks

Since reference inputs can be applied without causing windup effects (see the appendix) the problem of anti-windup control can be formulated in the following manner: *Design a compensator for desired disturbance rejection and assure that this compensator does not cause stability problems when sensor saturation occurs.* The standard solution to this problem known from the input-constrained case is to apply a stabilizing feedback of the difference between the saturated and the unsaturated signals. This would also solve the windup problem caused by saturating sensors (see Sect. 1.2). Unfortunately, this difference is not available.

The missing information problem was circumnavigated in Chaps. 1 and 2 by a switching element (the saturation indicator). It provides an information about the fact of saturation, but not about its extent. Nevertheless, this information is sufficient to realize a stabilizing feedback during saturation. Because the switching element is located directly behind the saturation nonlinearity all known methods for stability analysis of nonlinear systems indicate a failure of this approach. But all systems investigated so far have shown that it perfectly stabilizes the nonlinear loop in case of sensor saturation.

The missing proof of stability is not the only problem with the approach of Chap. 1. On the one hand, it is restricted to SISO systems and on the other hand, it entails an enormous blow-up of dimensions when it is used to solve the anti-windup problem for joint input and output saturation.

For the windup community, the main attraction of the method presented in Chap. 1 would be the fact that it follows the usual design paradigm of windup prevention (start with an arbitrary compensator, then apply anti-windup). Instead, one can lean back and pose the question: what is the primary objective in the presence of saturating sensors? The answer: find a means to obtain a desired amount of disturbance attenuation such that sensor saturation does not destabilize the loop. If one was able to design a disturbance rejecting control with desired performance such that the linear part of the loop satisfied the Circle Criterion (CC), windup would automatically be prevented and stability could be proven in a rigorous way!

© The Author(s), under exclusive license to Springer Nature Switzerland AG 2021
P. Hippe, *Windup in Control Owing to Sensor Saturation*,
SpringerBriefs in Applied Sciences and Technology,
https://doi.org/10.1007/978-3-030-73133-5_3

At first glance, this seems impossible because a compensator can either be designed for good disturbance attenuation or to obtain a linear part of the nonlinear loop satisfying the CC.

Actually, there is a way to get over this obstacle, namely, the namesake approach to feedback control, i.e., the **compensating approach**. When the poles of the plant are the zeros of the compensator and the zeros of the plant are the poles of the compensator (this is, of course, restricted to minimum-phase systems, the basic assumption also in Chap. 1), the remaining free eigenvalues of the closed loop can be shifted arbitrarily far into the left half s-plane without jeopardizing the stability of the nonlinear loop. And by a sufficient degree of left shift any rate of disturbance attenuation can be realized.

Though this does not follow the above-cited design paradigm, it has two advantages over the approach in Chap. 1. It allows a strict proof of stability and to solve both the SISO and the MIMO cases. Finally, it considerably simplifies the design of an anti-windup control in the presence of input *and* output saturation both for SISO and MIMO systems (see Chaps. 4 and 5).

3.2 Anti-windup Design for SISO Systems

We start with the frequency-domain description

$$y(s) = G(s)u(s) + G_d(s)d(s) \tag{3.1}$$

of the system, where $y(t) \in \mathbb{R}^1$ is the output, $u(t) \in \mathbb{R}^1$ is the manipulated input, and $d(t) \in \mathbb{R}^1$ is a persistent disturbance. The system is assumed to be strictly proper, asymptotically stable, and minimum phase. The transfer functions $G(s)$ and $G_d(s)$ are represented by

$$G(s) = \frac{N(s)}{D(s)}, \tag{3.2}$$

and

$$G_d(s) = \frac{N_d(s)}{D(s)}, \tag{3.3}$$

where $N(s)$ and $D(s)$ are coprime polynomials. The coefficient κ is the difference degree $\kappa \geq 1$.

It is assumed that the disturbance d can be modeled in a signal process with characteristic polynomial

$$C_d(s) = s^{n_d} + \psi_{n_d-1}s^{n_d-1} + \cdots + \psi_1 s + \psi_0, \tag{3.4}$$

and that $C_d(s)$ has only simple zeros on the imaginary axis and no zero in the open right half plane. This typically includes constant disturbances with $C_d(s) =$

s, sinusoidal disturbances with $C_d(s) = s^2 + \omega_0^2$, and sinusoidal disturbances with constant offset with $C_d(s) = s(s^2 + \omega_0^2)$. In order to reject such disturbances, the numerator polynomial $N(s)$ must not have zeros coinciding with the zeros of $C_d(s)$.

The measured output $y_s(t) \in \mathbb{R}^1$ is a saturated version of the output $y(t)$ of the system. The nonlinearity $y_s = \mathrm{sat}_{y_0}(y)$ is defined by (1.9).

The linear controller of the order n_C is supposed to be an observer-based compensator with internal model for robust disturbance rejection (see, e.g., [1, 2]). Its transfer behavior $u_C(s) = G_C(s)y_s(s)$ is characterized by

$$G_C(s) = \frac{N_C(s)}{D_C(s)}. \tag{3.5}$$

This compensator has been designed such that the linear loop with feedback interconnection $u = -u_C$ and $y_s = y$ is stable. The characteristic polynomial $C_P(s)$ of the linear loop can be represented as

$$C_P(s) = D_C(s)D(s) + N_C(s)N(s) = \Delta(s)\tilde{D}(s), \tag{3.6}$$

where $\Delta(s)$ is a Hurwitz polynomial of degree n_C characterizing the dynamics of the state-plus-disturbance observer and $\tilde{D}(s)$ is a Hurwitz polynomial of degree n characterizing the dynamics of the state feedback control [1, 2].

A robust rejection of the persistent disturbances results if the polynomial $D_C(s)$ has the form

$$D_C(s) = C_d(s)D_C^*(s). \tag{3.7}$$

It is assumed that $D_C^*(s)$ is a Hurwitz polynomial.

For the stability analysis of the loop, we again make use of the identity

$$\mathrm{sat}_{y_0}(y) = y - \mathrm{dead}_{y_0}(y), \tag{3.8}$$

where $\mathrm{dead}_{y_0}(y)$ is defined by (1.11). Therefore, the considered control loop has the form shown in Fig. 3.1. If the loop gain $G_{Ly}(s)$ in

$$y(s) = -G_{Ly}(s)\tilde{y}(s) \tag{3.9}$$

Fig. 3.1 Loop with modified representation of the saturating nonlinearity

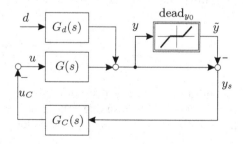

satisfies the CC, the nonlinear loop has a stable behavior, and if not, there is the danger of windup.

A compensator with internal signal model leading to a transfer function

$$G_{Ly}(s) = -\frac{N(s)N_C(s)}{\Delta(s)\tilde{D}(s)}, \tag{3.10}$$

(see also (1.13)) satisfying the CC usually leads to poor disturbance rejection. This will be demonstrated by a simple example.

Example 3.1 Consider a system of the order three with transfer functions

$$G(s) = \frac{N(s)}{D(s)} = \frac{2s+4}{3(s+1)^3} \quad \text{and} \quad G_d(s) = \frac{N_d(s)}{D(s)} = \frac{s^2+4s+6}{3(s+1)^3}. \tag{3.11}$$

Assume that step-like disturbances must be rejected so that the compensator has integral action (i.e., $C_d(s) = s$) and that an observer-based controller of minimum order is being used. Given the order $n = 3$ of the plant, one output $p = 1$, and the order $n_d = 1$ of the disturbance process, the order of the minimum-order compensator is $n_C = n - p + n_d = 3$. Choosing the same locations for the eigenvalues of the state feedback (zeros of the polynomial $\tilde{D}(s)$) and the observer (zeros of $\Delta(s)$) while compensating the zero at $s = -2$ in $\Delta(s)$, $G_{Ly}(s)$ satisfies the CC (i.e., the frequency response $G_{Ly}(j\omega)$ stays right of the point -1 in the complex plane) for $\tilde{D}(s) = 3(s+1.5)^3$ and $\Delta(s) = (s+2)(s+1.5)^2$, leading to a compensator with transfer function

$$G_C(s) = \frac{N_C(s)}{D_C(s)} = \frac{9s^3 + 28.875s^2 + 31.21875s + 11.390625}{s^3 + 6.5s^2 + 9s}. \tag{3.12}$$

For eigenvalues further in the left half s-plane, the CC is not satisfied. Applying a unit step input $d(t) = 1(t)$ the disturbance transient in the full line in Fig. 3.2 results. This has two drawbacks. On the one hand, the disturbance rejection is quite poor. On the other hand, an existing sensor saturation could already become active for small disturbance inputs. Though this does not cause an unstable behavior, it opens the loop and leads to prolonged disturbance transients and to a considerably increased output $y(t)$.

Shifting the eigenvalues further left leads to a better disturbance rejection. For $\tilde{D}(s) = 3(s+4)^3$ and $\Delta(s) = (s+2)(s+4)^2$, the compensator obtains the form

$$G_C(s) = \frac{N_C(s)}{D_C(s)} = \frac{159s^3 + 882s^2 + 1894.5s + 1536}{s^3 + 19s^2 + 34s}. \tag{3.13}$$

The disturbance step response resulting with this compensator is shown in dotted lines in Fig. 3.2. The resulting transfer function $G_{Ly}(s)$, however, violates the CC so that stability of the nonlinear loop is no longer guaranteed.

Fig. 3.2 Disturbance step responses

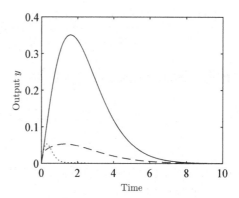

A good disturbance attenuation with guaranteed stability of the nonlinear loop can be obtained when applying the *compensating approach*, i.e., when compensating the dynamics of the plant by the compensator. In the observer-based design, this can be achieved by choosing $\tilde{D}(s) = D(s)$ and $\Delta(s) = \Delta_r(s)N(s)$ with $\Delta_r(s) = (s + \alpha)^\kappa$. For all $\alpha > 0$, the resulting transfer function $G_{Ly}(s)$ satisfies the CC.

The same reduced peak value as with compensator (3.13) results for $\alpha = 21$, i.e., for $\Delta_r(s) = (s + 21)^2$, leading to

$$G_C(s) = \frac{441 D(s)}{s(s + 42)N(s)}. \tag{3.14}$$

The corresponding $G_{Ly}(s)$ has the form $G_{Ly}(s) = -441/(s + 21)^2$ which satisfies the CC. The broken line in Fig. 3.2 shows the disturbance step response resulting with compensator (3.14).

The best disturbance rejection with guaranteed stability always results when the two polynomials $\tilde{D}(s)$ and $D(s)$ coincide and when the zeros of the system are also zeros of the observer polynomial $\Delta(s)$.

For step-like disturbances (i.e., $C_d(s) = s$) and when the κ (difference degree) eigenvalues of $\Delta(s)$ are placed to identical locations $s = -\alpha, \alpha > 0$, the transfer function $G_{Ly}(s)$ obtains the form

$$G_{Ly}(s) = -\frac{\alpha^\kappa}{(s + \alpha)^\kappa},$$

and this satisfies the CC for all $\alpha > 0$.

Given a system and a persistent disturbance modeled by a process with characteristic polynomial $C_d(s)$ the design of the compensator with

$$\tilde{D}(s) = D(s) \quad \text{and} \quad \Delta(s) = \Delta_r(s)N(s) \tag{3.15}$$

can be carried out in the following way.

First decompose $\Delta_r(s)$ according to

$$\Delta_r(s) = Q_D(s)C_d(s) + R_D(s). \tag{3.16}$$

Then choose the compensator polynomials as

$$N_C(s) = R_D(s)D(s), \tag{3.17}$$

$$D_C(s) = C_d(s)Q_D(s)N(s). \tag{3.18}$$

With this compensator, the characteristic polynomial (3.6) has the desired zeros and the loop gain becomes

$$G_{Ly}^{comp}(s) = -\frac{R_D(s)}{\Delta_r(s)}. \tag{3.19}$$

For a disturbance process with $C_p(s) = s$ (i.e., for a compensator with integral action), the transfer function (3.19) has the special form

$$G_{Ly}^{comp}(s) = -\frac{\Delta_r(0)}{\Delta_r(s)}. \tag{3.20}$$

Thus, for arbitrary polynomials $\Delta_r(s)$ with real zeros in the left half s-plane, the transfer functions (3.19), (3.20) satisfy the CC. By placing the zeros of $\Delta_r(s)$ far left the peak amplitude of the disturbance step response can be reduced at will.

Example 3.2 In Example 1.2, a system with four poles and two zeros is investigated. Its transfer functions have the form

$$G(s) = \frac{6.25s^2 + 37.5s + 50}{s^4 + 8s^3 + 39s^2 + 62s + 50},$$

and

$$G_d(s) = \frac{20s^3 + 80s^2 + 170s + 300}{s^4 + 8s^3 + 39s^2 + 62s + 50}.$$

It is assumed that a step-like disturbance $d(t) = 1(t)$ enters the system (i.e., $C_d(s) = s$). The sensor saturates at $y_0 = 0.2$. In Example 1.2, an "arbitrary" compensator with integral action was designed placing two poles below the zeros and the remaining ones to $s = -20$. The resulting linear part of the loop does not satisfy the CC. It is so badly violated that the loop starts limit cycling with the above output saturation limit. In Example 1.2, this windup is prevented by a modification of the numerator of the compensator with the aid of the saturation indicator. This leads to the disturbance transients in Fig. 1.5.

Using the compensating design one can achieve the same amount of disturbance attenuation. With $\tilde{D}(s) = D(s)$ and $\Delta_r(s) = (s + 105)^2$ the above-described compensator design yields the transfer function

Fig. 3.3 Disturbance
transients with $\alpha = 105$

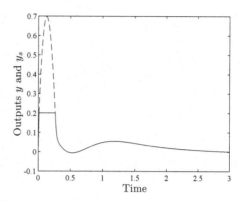

$$G_C(s) = \frac{11025 D(s)}{(s^2 + 210s)N(s)},$$

and with this compensator the linear part of the loop

$$G_{Ly}(s) = \frac{11025}{(s + 105)^2}$$

satisfies the CC. The zeros of $\Delta_r(s)$ were chosen to obtain the same disturbance attenuation as in Example 1.2. Figure 3.3 shows the disturbance step responses with the above compensator. The full line depicts the output $y_s(t)$ and the broken line the output $y(t)$. The amplitude of $y(t)$ is the same as in Fig. 1.5, but the transients in both figures are not coinciding because the eigenvalues of the system are now zeros of the characteristic polynomial $C_P(s)$ of the loop.

In Example 1.2, the closed-loop poles were shifted further left to obtain a disturbance transient not exceeding the sensor saturation limit. This is also feasible with the compensating approach. Choosing $\tilde{D}(s) = D(s)$ and $\Delta_r(s) = (s + 180)^2$ the compensator results as

$$G(s) = \frac{32400 D(s)}{(s^2 + 360s)N(s)},$$

and, of course, the CC is satisfied.

Figure 3.4 shows the output $y = y_s$ and the input u due to a disturbance step input $d(t) = 1(t)$. As in Example 1.2, the output saturation does not become active and the amplitude of the input signal is also about the same.

This shows that one can design a "compensating" control with integral action for an arbitrarily good disturbance attenuation without jeopardizing the stability of the nonlinear loop.

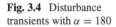

Fig. 3.4 Disturbance transients with $\alpha = 180$

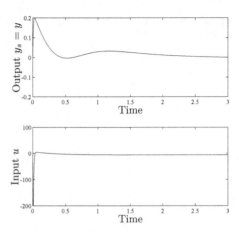

Table 3.1 Zeros of $\Delta_r(s)$ assuring stability for sinusoidal disturbances with $\omega_0 = 1$

$\kappa + 1$	s_1	s_2	s_3	s_4	s_5
2	10	0.1			
3	15	15	0.10026		
4	20	20	20	0.10033	
5	20	20	20	20	0.12571

However, this left shift of the zeros can lead to enormous amplitudes of the controller output signal, thus violating the always-existing input restrictions. When choosing

$$\Delta_r(s) = (s + \alpha)^\kappa, \tag{3.21}$$

the control design depends on a single parameter $\alpha > 0$. This parameter can be used to assign the desired behavior, either in view of the disturbance attenuation or the resulting control signal amplitude. If it cannot be ruled out that the control signal exceeds the always-existing limitations the problem of joint input and output saturation has to be considered (see Chap. 4).

For signal models of higher orders than one, the zeros of $\Delta_r(s)$ can no longer be placed arbitrarily. Instead, appropriate locations for these zeros have to be determined. There exist infinitely many solutions so that a "best" solution cannot be presented. In the sequel, possible locations for the zeros of $\Delta_r(s)$ are listed.

Given a SISO system of the order n and sinusoidal disturbances modeled by $C_d(s) = s^2 + \omega_0^2$. Then the order of the minimum-order state-plus-disturbance observer is $n_C = n - 1 + 2 = n + 1$. Table 3.1 shows the zeros $s_i = -\alpha_i$ of $\Delta_r(s)$ for different degrees $\kappa + 1$ which assure that (3.19) satisfies the CC for disturbances modeled by $C_d(s) = s^2 + 1$.

Table 3.2 Zeros of $\Delta_r(s)$ assuring stability for disturbances modeled by $C_d(s) = s(s^2 + 1)$

$\kappa + 2$	s_1	s_2	s_3	s_4	s_5	s_6
3	4.5	0.10977221	s_2			
4	6.5	s_1	0.11543939	s_3		
5	10	s_1	s_1	0.10033279	s_4	
6	12	s_1	s_1	s_1	0.104664952	s_5

The polynomials defined in Table 3.1 only assure stability of the nonlinear loop for sinusoidal disturbances with $\omega_0 = 1$. For all other sinusoidal disturbances, different sets of parameters are valid. However, the above numbers are also useful for $\omega_0 \neq 1$.

The rationale for this is the following. One can always carry out a time scaling of the system to be controlled such that the attacking disturbance has the scaled frequency $\omega_{0s} = 1$. Since the plant dynamics are totally compensated by the controller, the numbers in Table 3.1 assure that $G_{Ly}(s)$ satisfies the CC for the scaled system. To obtain the results for the original system, the time scaling has to be reversed, which also has to be carried out for the polynomial $\Delta_r(s)$. Therefore, one obtains the corresponding polynomial $\Delta_r(s)$ for $\omega_0 \neq 1$ simply by multiplying the numbers of Table 3.1 by ω_0.

One could, of course, also find suitable polynomials $\Delta_r(s)$ having the form $\Delta_r(s) = (s + \alpha)^{(\kappa+1)}$. The zeros in Table 3.1, however, assure a smaller maximum amplitude of the disturbance transients and thus, for a given amplitude of the attacking disturbances, better prevent an activation of sensor saturation.

Given a SISO system of the order n and sinusoidal disturbances with constant offset modeled by $C_d(s) = s(s^2 + \omega_0^2)$. Then the order of the minimum-order state-plus-disturbance observer is $n_C = n - 1 + 3 = n + 2$. Table 3.2 shows the zeros $s_i = -\alpha_i$ for $\Delta_r(s)$ of different degrees $\kappa + 2$ which assure that (3.19) satisfies the CC for disturbances modeled by $C_d(s) = s(s^2 + 1)$.

Also here, the zeros of $\Delta_r(s)$ characterize the compensator in the presence of a sinusoidal part with $\omega_0 = 1$. The zeros of $\Delta_r(s)$ for disturbances with $\omega_0 \neq 1$ can be obtained by multiplying the numbers in Table 3.2 by ω_0.

For very low frequencies, the dynamics of the closed loop become quite slow. Then, one could either try to find "better" polynomials $\Delta_r(s)$ or use a compensator with integral action, which can be tuned to obtain nearly ideal disturbance attenuation.

A seeming drawback of the presented approach is the compensation of the system dynamics. This can lead to poorly damped transients of the closed loop if the system has poorly damped poles. It should be kept in mind, however, that it will always be a hard problem to suppress disturbances in a poorly damped system when sensor saturation occurs.

3.3 Résumé of the SISO Design Steps

We start with a minimum-phase, strictly proper system of the order n, described by its transfer function

$$G(s) = \frac{N(s)}{D(s)}. \tag{3.22}$$

The difference degree of this system is κ. The observer-based compensator with transfer function

$$G_C(s) = \frac{N_C(s)}{D_C(s)} \tag{3.23}$$

is designed to reject persistent disturbances modeled in a signal process with characteristic polynomial

$$C_d(s) = s^{n_d} + \psi_{n_d-1}s^{n_d-1} + \cdots + \psi_1 s + \psi_0, \tag{3.24}$$

so that its denominator has the form

$$D_C(s) = C_d(s)D_C^*(s). \tag{3.25}$$

Choose $\tilde{D}(s) = D(s)$ and an observer polynomial $\Delta(s) = \Delta_r(s)N(s)$ with $\Delta_r(s)$ a polynomial of degree $\kappa + n_d - 1$.
    Now decompose $\Delta_r(s)$ according to

$$\Delta_r(s) = Q_D(s)C_d(s) + R_D(s). \tag{3.26}$$

Then the compensator polynomials have the forms

$$N_C(s) = R_D(s)D(s), \tag{3.27}$$

and

$$D_C(s) = C_d(s)Q_D(s)N(s). \tag{3.28}$$

For $n_d = 1$, i.e., for step-like disturbances, the zeros of $\Delta_r(s) = (s + \alpha)^\kappa$ can be chosen freely to obtain a desired disturbance attenuation with guaranteed stability.
    For $n_d > 1$, the zeros of $\Delta_r(s)$ must be chosen such that the transfer function

$$G_{Ly}^{comp}(s) = -\frac{R_D(s)}{\Delta_r(s)} \tag{3.29}$$

satisfies the CC. Possible zeros are listed in Tables 3.1 and 3.2.

Example 3.3 To demonstrate the application of the information contained in Table 3.2 consider the system already used in Example 3.1 with its transfer functions

$$G(s) = \frac{N(s)}{D(s)} = \frac{2s+4}{3(s+1)^3},$$

and

$$G_d(s) = \frac{N_d(s)}{D(s)} = \frac{s^2 + 4s + 6}{3(s+1)^3}.$$

We now assume that a sensor saturation limit at $y_0 = 1$ exists and a disturbance consisting of a sine signal with constant offset enters the system. The oscillating part has a frequency of $\omega_0 = 3$ so that the characteristic polynomial of the disturbance signal model has the form $C_d(s) = s(s^2 + 9)$. Since the difference degree of the system is $\kappa = 2$, the polynomial $\Delta_r(s)$ is of degree 4 and the zeros of this polynomial assuring stability are listed in the second line of Table 3.2 for $\omega_0 = 1$.

With $\omega_0 = 3$ the polynomial $\Delta_r(s)$ obtains the form

$$\Delta_r(s) = (s + 19.5)^2 (s + 0.34632)^2.$$

The decomposition (3.16) or (3.26) yields

$$Q_D(s) = s + 39.69264,$$

and

$$R_D(s) = 398.3829s^2 - 89.17984s + 45.60625.$$

Inserting $R_D(s)$ in (3.17) or in (3.27) gives the numerator polynomial

$$N_C(s) = (398.3829s^2 - 89.17984s + 45.60625)D(s),$$

and inserting $Q_D(s)$ in (3.18) or in (3.28) yields the denominator polynomial

$$D_C(s) = (s^3 + 9s)(s + 39.69264)N(s)$$

of the compensator. With this compensator the frequency response of the linear part of the loop

$$G_{Ly}(j\omega) = -\frac{N(j\omega)N_C(j\omega)}{C_P(j\omega)}$$

(see (3.10)) has the form shown in Fig. 3.5. It stays right of the point $(-1, j0)$ so that the CC is satisfied, and consequently sensor saturation does not trigger windup effects.

Figure 3.6 shows the reactions of the nonlinear loop to an input disturbance $d(t) = 1(t) + \sin(3t)$ when an output limitation with $y_0 = 1$ is present.

Fig. 3.5 Frequency
response $G_{Ly}(j\omega)$

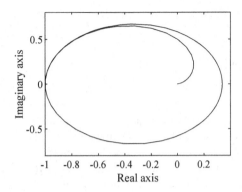

Fig. 3.6 Disturbance
transients of the example

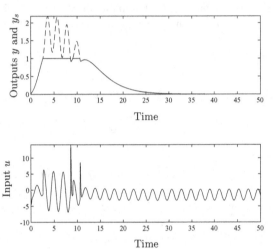

The upper part shows in broken lines the output $y(t)$ and in full lines the output $y_s(t)$ of the sensor. The lower part of Fig. 3.6 depicts the input $u(t)$. The attenuation of the sinusoidal part of the disturbance is not possible during the first 8 s because the input signal y_s to the compensator is constant during this period.

3.4 Anti-windup Design for MIMO Systems

The compensating design for SISO systems described in Sect. 3.2 has two advantages. It allows a strict proof of stability and it is very easy to apply, because the compensator with integral action is parametrized by one parameter α only and its design simply requires the decomposition of a polynomial. For more complex disturbance signal models, the parametrization is slightly more complicated, but the design equations remain unchanged. It turns out that, when using the frequency-domain approach on the basis of coprime Matrix Fraction Descriptions (MFDs), the anti-windup design for MIMO systems in the presence of sensor saturation is nearly as simple as for SISO systems.

We consider strictly proper, asymptotically stable, completely controllable and observable minimum-phase LTI systems of the order n with manipulated input $u(t) \in \mathbb{R}^p$ and output $y(t) \in \mathbb{R}^p$. The transfer behavior of these systems is described by

$$y(s) = G(s)u(s), \tag{3.30}$$

where $G(s)$ is either represented by its right coprime MFD

$$G(s) = N(s)D^{-1}(s), \tag{3.31}$$

or by its left coprime MFD

$$G(s) = \bar{D}^{-1}(s)\bar{N}(s), \tag{3.32}$$

with $D(s)$ column reduced and both $\bar{N}(s)$ and $\bar{D}(s)$ row reduced. The coefficient κ is defined as the maximum difference degree of all outputs, namely,

$$\kappa = \overset{\max}{j} \left\{ \delta_{rj}[\bar{D}(s)] - \delta_{rj}[\bar{N}(s)] \right\}, \tag{3.33}$$

where $\delta_{rj}[\cdot]$ is the jth row degree. It is further assumed that persistent disturbances $d(t) \in \mathbb{R}^q$ affect the system and that the transfer behavior between these disturbances and the outputs is described by

$$y(s) = G_d(s)d(s) = \bar{D}^{-1}(s)\bar{N}_d(s)d(s), \tag{3.34}$$

with $\bar{D}(s)$ as in (3.32). The q disturbances can be modeled in a signal process with characteristic polynomial $C_d(s)$ (see (3.4)). And the disturbances can only be rejected if the polynomial matrix $N(s)$ does not have transmission zeros coinciding with zeros of $C_d(s)$.

The measured output vector $y_s(t) \in \mathbb{R}^p$ is a saturated version of the output y of the system. The components of the nonlinearity $y_s = \text{sat}_{y_0}(y)$ are defined by

$$\text{sat}_{y_0}(y_i) := \begin{cases} y_{0i} & \text{if } y_i > y_{0i} > 0 \\ y_i & \text{if } -y_{0i} \le y_i \le y_{0i}; \quad i = 1, 2, \ldots, p . \\ -y_{0i} & \text{if } y_i < -y_{0i} \end{cases} \tag{3.35}$$

The linear controller of the order n_C is supposed to be an observer-based compensator with internal model for robust disturbance rejection (see, e.g., [2]). Its transfer behavior is described by

$$u_C(s) = G_C(s)y_s(s), \tag{3.36}$$

where $G_C(s)$ is represented by its left coprime MFD

$$G_C(s) = D_C^{-1}(s)N_C(s), \qquad (3.37)$$

with $D_C(s)$ row reduced. This compensator has been designed such that the closed loop with feedback interconnection $u = -u_C$ and $y_s = y$ is stable.

The characteristic polynomial matrix $C_P(s)$ of the linear loop can be represented according to

$$C_P(s) = D_C(s)D(s) + N_C(s)N(s) = \Delta(s)\tilde{D}(s), \qquad (3.38)$$

where the $p \times p$ polynomial matrix $\Delta(s)$ characterizes the dynamics of the state-plus-disturbance observer (the zeros of det $\Delta(s)$ are the eigenvalues of the observer) and the $p \times p$ polynomial matrix $\tilde{D}(s)$ characterizes the state feedback control [2].

For simplicity, it is assumed that the q attacking disturbances have the same signal characteristic (3.4). Define the $p \times p$ polynomial matrix

$$D_{dis}(s) = \text{diag}[C_d(s)]. \qquad (3.39)$$

Then, a robust disturbance rejection results when the denominator matrix $D_C(s)$ of the compensator has the form

$$D_C(s) = D_C^*(s)D_{dis}(s). \qquad (3.40)$$

It is supposed that det $D_C^*(s)$ is a Hurwitz polynomial. The design of such compensators is facilitated (in analogy to the SISO case) when $\bar{N}(s)$ is a right divisor of $\Delta(s)$.

The results for SISO systems in Sect. 3.2 can be used to develop the anti-windup solution for MIMO systems. The first step in the compensating design is the choice of

$$\tilde{D}(s) = D(s). \qquad (3.41)$$

For $\tilde{D}(s) = D(s)$, Equation (3.38) only has a solution if $D(s)$ is a right divisor of $N_C(s)N(s)$. Observing the fact

$$\bar{D}(s)N(s) = \bar{N}(s)D(s), \qquad (3.42)$$

(compare (3.31) and (3.32)) it becomes obvious that $D(s)$ is a right divisor of $N_C(s)N(s)$ if $N_C(s)$ has the form

$$N_C(s) = N_{Cr}(s)\bar{D}(s), \qquad (3.43)$$

where $N_{Cr}(s)$ is a $p \times p$ polynomial matrix. The polynomial matrix $D(s)$ can thus be extracted from (3.38) giving the reduced form

$$D_C(s) + N_{Cr}(s)\bar{N}(s) = \Delta(s). \qquad (3.44)$$

Since $\bar{N}(s)$ is row reduced one can choose

$$D_C(s) = D_{Cr}(s)\bar{N}(s), \qquad (3.45)$$

and

$$\Delta(s) = \Delta_r(s)\bar{N}(s), \qquad (3.46)$$

so that also the zeros of the system can be extracted from (3.44) and consequently also from (3.38) giving the completely reduced design equation

$$D_{Cr}(s) + N_{Cr}(s) = \Delta_r(s) \qquad (3.47)$$

of the compensator. Now choose the $p \times p$ polynomial matrix $\Delta_r(s)$ as

$$\Delta_r(s) = \text{diag}[\hat{\Delta}_r(s)], \qquad (3.48)$$

with the polynomial $\hat{\Delta}_r(s)$ of the degree $\kappa - 1 + n_d$. For step-like disturbances (i.e., for $C_d(s) = s$), this polynomial can be chosen as

$$\hat{\Delta}_r(s) = (s + \alpha)^\kappa, \qquad (3.49)$$

with α an arbitrary positive number and for disturbance processes of the order $n_d > 1$ possible zeros of $\hat{\Delta}_r(s)$ are listed in Tables 3.1 and 3.2. Now with the decomposition

$$\Delta_r(s) = Q_D(s)D_{dis}(s) + R_D(s), \qquad (3.50)$$

the polynomial matrices of the compensator have the forms

$$N_C(s) = R_D(s)\bar{D}(s), \qquad (3.51)$$

and

$$D_C(s) = Q_D(s)\bar{N}(s)D_{dis}(s). \qquad (3.52)$$

This compensator obviously assures a robust rejection of the modeled disturbances and by construction it assigns the dynamics of the linear loop defined by the polynomial matrices (3.41) and (3.46).

The block diagram of the loop considered is again the one of Fig. 3.1. Now, however, the entities $G(s)$, $G_a(s)$, and $G_C(s)$ are transfer matrices and since y and $\tilde{y}$ are vectors, the saturating nonlinearity in $\tilde{y} = \text{dead}_{y_0}(y)$ is a diagonal matrix whose entries are defined by

$$\text{dead}_{y_0}(y_i) := \begin{cases} y_i - y_{0i} & \text{if} \quad y_i > y_{0i} > 0 \\ 0 & \text{if} \quad -y_{0i} \le y_i \le y_{0i}; \quad i = 1, 2, \ldots, p \;. \\ y_i + y_{0i} & \text{if} \quad y_i < -y_{0i} \end{cases} \qquad (3.53)$$

With the above-designed compensator and $\tilde{D}(s) = D(s)$, the matrix $G_{Ly}(s)$ in

$$y(s) = -G_{Ly}(s)\tilde{y}(s) \tag{3.54}$$

obtains the form

$$G_{Ly}(s) = -N(s)D^{-1}(s)\Delta^{-1}(s)N_C(s), \tag{3.55}$$

and with (3.31), (3.32), (3.46) and (3.51) this becomes

$$G_{Ly}(s) = -\bar{D}^{-1}(s)\bar{N}(s)\bar{N}^{-1}(s)\Delta_r^{-1}(s)R_D(s)\bar{D}(s). \tag{3.56}$$

And since $\Delta_r(s)$ and $R_D(s)$ are diagonal matrices with equal entries, one finally obtains

$$G_{Ly}(s) = -\Delta_r^{-1}(s)R_D(s), \tag{3.57}$$

which is also diagonal with equal entries of the form

$$G_{Ly}^{ii}(s) = -\frac{R_D^{ii}(s)}{\hat{\Delta}_r(s)}, \quad i = 1, 2, \ldots, p. \tag{3.58}$$

These transfer functions on the main diagonal coincide with (3.19) obtained for SISO systems. And these transfer functions have been shown to satisfy the CC.

Thus, the problem of windup prevention for MIMO systems in the presence of sensor saturation has a solution analogous to the results developed for SISO systems.

3.5 Résumé of the MIMO Design Steps

We start with a strictly proper, asymptotically stable, minimum-phase $p \times p$ MIMO system of the order n, described by its transfer matrix

$$G(s) = N(s)D^{-1}(s) = \bar{D}^{-1}(s)\bar{N}(s), \tag{3.59}$$

where $D(s)$ is column reduced and both $\bar{D}(s)$ and $\bar{N}(s)$ are row reduced. The maximum difference degree κ of all outputs is defined by (3.33). The observer-based compensator with its transfer matrix

$$G_C(s) = D_C^{-1}(s)N_C(s) \tag{3.60}$$

is designed to reject persistent disturbances modeled in a signal process with characteristic polynomial $C_d(s)$ (see (3.4)), so that its denominator matrix has the form

$$D_C(s) = D_C^*(s)D_{dis}(s) \tag{3.61}$$

with

$$D_{dis}(s) = \text{diag}[C_d(s)]. \tag{3.62}$$

For the compensating design, choose $\tilde{D}(s) = D(s)$ and the polynomial matrix $\Delta(s) = \Delta_r(s)\tilde{N}(s)$. The $p \times p$ polynomial matrix $\Delta_r(s)$ has the form

$$\Delta_r(s) = \text{diag}[\hat{\Delta}_r(s)], \tag{3.63}$$

with the polynomial $\hat{\Delta}_r(s)$ of the degree $\kappa - 1 + n_d$. For step-like disturbances (i.e., for $C_d(s) = s$), this polynomial can be chosen as

$$\hat{\Delta}_r(s) = (s + \alpha)^\kappa, \tag{3.64}$$

with α an arbitrary positive number and for disturbance processes of the order $n_d > 1$ possible zeros of $\hat{\Delta}_r(s)$ are listed in Tables 3.1 and 3.2. Now decompose

$$\Delta_r(s) = Q_D(s)D_{dis}(s) + R_D(s). \tag{3.65}$$

Then the polynomial matrices of the compensator have the forms

$$N_C(s) = R_D(s)\bar{D}(s), \tag{3.66}$$

and

$$D_C(s) = Q_D(s)\tilde{N}(s)D_{dis}(s). \tag{3.67}$$

Example 3.4 As a demonstrating example, consider a system with three inputs and outputs whose transfer behavior $G(s) = N(s)D^{-1}(s) = \bar{D}^{-1}(s)\tilde{N}(s)$ is characterized by

$$N(s) = \begin{bmatrix} 15s + 12 & 2 & -3 \\ 5s - 1 & 1 & 0 \\ s + 6 & 0 & 1 \end{bmatrix}, \quad D(s) = \begin{bmatrix} -s^3 + 3s^2 + 9s + 5 & s + 1 & s^2 + s - 1 \\ -2s - 4 & 0 & -s^2 - 3s \\ -2s^3 - 2s^2 - s - 3 & s + 1 & s^2 + s + 1 \end{bmatrix},$$

$$\bar{D}(s) = \begin{bmatrix} s^2 + 6.25s + 6.5 & -2.5s - 5 & -0.25s - 2.5 \\ -0.5s^2 + 0.625s + 1.25 & s^2 + 1.75s + 0.5 & 0.375s - 0.25 \\ -4.25s - 4.5 & 2.5s + 3 & s^2 + 2.25s + 2.5 \end{bmatrix},$$

and

$$\bar{N}(s) = \begin{bmatrix} 4s + 15 & 2s + 7 & -2s - 7 \\ 3.5 & 1.5 & -0.5 \\ -11 & -7 & 5 \end{bmatrix}.$$

The maximum difference degree of all outputs is $\kappa = 2$. The eigenvalues of the system are located at $s_{1,2,3} = -1$ and $s_{4,5,6} = -2$ and the transmission zero at $s = -4$. Three step-like input disturbances affect the system via

$$\bar{N}_d(s) = \begin{bmatrix} 2s + 4 & 6 & 2 \\ 2 & 4s + 4 & -s - 2 \\ -2 & -s - 2 & s + 6 \end{bmatrix},$$

so that one needs a compensator with integral action. Consequently, the polynomial matrix $D_{dis}(s)$ has the form

$$D_{dis}(s) = \begin{bmatrix} s & 0 & 0 \\ 0 & s & 0 \\ 0 & 0 & s \end{bmatrix}.$$

Assume the disturbances are $d_1(t) = d_{s1}1(t)$, $d_2(t) = d_{s2}1(t)$, and $d_3(t) = d_{s3}1(t)$ with $d_{s1} = 2$, $d_{s2} = 2$, and $d_{s3} = -2$ and the sensors saturate at $y_{01} = 1$, $y_{02} = 1.5$, and $y_{03} = 1$.

Using a "slow" compensator on the basis of $\hat{A}_r = (s + 3)^2$ the compensating approach yields the polynomial matrices

$$D_C(s) = \begin{bmatrix} s^2 + 6s & 0 & 0 \\ 0 & s^2 + 6s & 0 \\ 0 & 0 & s^2 + 6s \end{bmatrix} \bar{N}(s),$$

and

$$N_C(s) = \begin{bmatrix} 9 & 0 & 0 \\ 0 & 9 & 0 \\ 0 & 0 & 9 \end{bmatrix} \bar{D}(s)$$

of the compensator. The diagonal entries (3.58) of the transfer matrix (3.57) have the form

$$G_{Ly}^{ii}(s) = -\frac{9}{(s + 3)^2}, \quad i = 1, 2, 3,$$

so that $G_{Ly}(s)$ satisfies the CC and the stability of the nonlinear loop is guaranteed. The left-hand side of Fig. 3.7 shows the disturbance transients of the MIMO system. The full lines depict the output y_s of the saturating sensors and the broken lines the output y of the system. Also shown are the input signals generated by the above compensator.

It is obvious that the real outputs of the system exceed the outputs of the sensors considerably. It thus can be desirable to obtain a better attenuation of the disturbances to assure that the sensor saturation does not become active for the expected disturbances.

If, instead of the above $\hat{A}_r(s)$, one chooses another design polynomial, namely, $\hat{A}_r = (s + 20)^2$ a compensator with

$$D_C(s) = \begin{bmatrix} s^2 + 40s & 0 & 0 \\ 0 & s^2 + 40s & 0 \\ 0 & 0 & s^2 + 40s \end{bmatrix} \bar{N}(s),$$

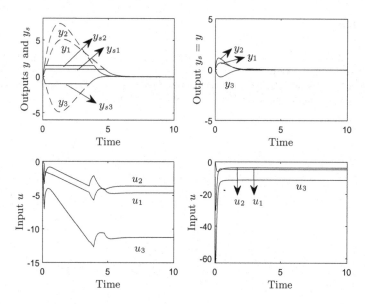

Fig. 3.7 Disturbance transients of the example

and

$$N_C(s) = \begin{bmatrix} 400 & 0 & 0 \\ 0 & 400 & 0 \\ 0 & 0 & 400 \end{bmatrix} \bar{D}(s)$$

results. Then, the diagonal entries (3.58) of the transfer matrix (3.57) have the form

$$G_{Ly}^{ii}(s) = -\frac{400}{(s+20)^2}, \quad i = 1, 2, 3,$$

so that $G_{Ly}(s)$ again satisfies the CC and stability is guaranteed.

The right-hand side of Fig. 3.7 shows the disturbance transients resulting with the fast compensator. The upper part depicts the outputs y and y_s which now coincide because sensor saturation does not occur. The lower part shows the corresponding plant input signal generated by this fast compensator. It is obvious that the improved disturbance attenuation is bought by enormous input signals so that the always-existing input signal limitations could be exceeded.

This would necessitate the design of a compensator guaranteeing stability in the joint presence of input and output saturation. Such a compensator will be much slower (see Chap. 5). It therefore can be advantageous to use the parameter α in a way that the best disturbance attenuation takes place under the condition that the control signal does not saturate.

References

1. P. Hippe, *Windup in Control—Its Effects and Their Prevention* (Springer, Berlin, Heidelberg, New York, London, 2006)
2. P. Hippe, J. Deutscher, *Design of Observer-based Compensators—From the Time to the Frequency Domain* (Springer, Berlin, Heidelberg, New York, London, 2009)

Chapter 4
Windup Prevention in the Joint Presence of Input and Output Saturation—The SISO Case

4.1 Introductory Remarks

Here, and in the next chapter, we restrict out attention to compensators with integral action. The best results for disturbance rejection in the presence of saturating sensors are, of course, achievable when the danger of an additional input saturation is not present or *vice versa*. In view of unknown external disturbances, this cannot always be guaranteed.

At the latest when solving the problems arising from input *and* output constraints, one has to quit the classic design paradigm. This became already clear in Chap. 2 when this problem was solved with the aid of the saturation indicator. Unfortunately, the equivalent Additional Dynamic Element (ADE), which replaced the saturation indicator in the stability studies, increased the order of the system for the test of the Circle Criterion (CC) by a factor two.

This increase in order does not occur when solving the sensor saturation problem with the aid of the compensating approach (see Chap. 3), i.e., when abandoning the classic design paradigm from the outset. The compensating approach additionally simplifies the compensator design. As we are considering constant or step-like changing disturbances, the parametrization of the compensator depends on one parameter α only.

This considerably simplifies the solution of the anti-windup problem in the presence of joint input and output constraints. In Chap. 2, the n roots of $\tilde{D}(s)$ and the κ roots of $\Delta(s)$ had to be modified in an appropriate manner during the trial-and-error search for the best polynomial $\Delta_1(s)$. When using the compensating approach, only one parameter α needs to be adapted in this process.

The anti-windup control problem for input- and output-constrained SISO systems can nearly always be solved. For some systems, however, it is impossible or a solution only exists with an extreme low-gain compensator.

In the MIMO case to be discussed in Chap. 5, additional noxious effects will arise.

© The Author(s), under exclusive license to Springer Nature Switzerland AG 2021
P. Hippe, *Windup in Control Owing to Sensor Saturation*,
SpringerBriefs in Applied Sciences and Technology,
https://doi.org/10.1007/978-3-030-73133-5_4

4.2   Problem Formulation

We consider asymptotically stable, strictly proper, minimum-phase, completely controllable, and observable LTI systems of the order n with one input u_s and one output y. The transfer behavior of these systems is described by

$$y(s) = G(s)u_s(s). \tag{4.1}$$

It is further assumed that a constant or step-like changing disturbance d affects the system and that the transfer behavior between this disturbance and the output y is described by

$$y(s) = G_d(s)d(s). \tag{4.2}$$

The transfer function $G(s)$ is represented as

$$G(s) = \frac{N(s)}{D(s)}, \tag{4.3}$$

with $N(s)$ and $D(s)$ coprime Hurwitz polynomials where, for simplicity, it is assumed that $D(s)$ is a monic polynomial.

The transfer function $G_d(s)$ is represented as

$$G_d(s) = \frac{N_d(s)}{D(s)}, \tag{4.4}$$

with $D(s)$ as defined in (4.3). Since $N(0) \neq 0$ constant disturbances can be rejected.

The linear compensator is described by

$$u_C(s) = G_C(s)y_s(s), \tag{4.5}$$

where y_s is the output of the sensor. This compensator has been designed as an observer-based compensator for robust rejection of constant disturbances such that the closed loop with feedback interconnection $u_s = -u_C$ and $y_s = y$ is stable. The transfer function $G_C(s)$ is represented by

$$G_C(s) = \frac{N_C(s)}{D_C(s)}. \tag{4.6}$$

The compensator contains integral action so that one has

$$D_C(s) = s D_C^*(s). \tag{4.7}$$

It is assumed that a state observer of minimum order is being used so that the order of the compensator with integral action is $n_C = n$ and we also assume that $D_C^*(s)$ is a Hurwitz polynomial.

The input $u_s = \text{sat}_{u_0}(u)$ to the system is a saturated version of the command signal

$$u = -u_C. \tag{4.8}$$

The nonlinearity $u_s = \text{sat}_{u_0}(u)$ is defined by

$$\text{sat}_{u_0}(u) := \begin{cases} u_0 & \text{if} \quad u > u_0 > 0 \\ u & \text{if} \quad -u_0 \le u \le u_0 \\ -u_0 & \text{if} \quad u < -u_0. \end{cases} \tag{4.9}$$

The measured output $y_s(t)$ is a saturated version of the output y of the system. The nonlinearity $y_s = \text{sat}_y(y)$ is defined by

$$\text{sat}_{y_0}(y) := \begin{cases} y_0 & \text{if} \quad y > y_0 > 0 \\ y & \text{if} \quad -y_0 \le y \le y_0 \\ -y_0 & \text{if} \quad y < -y_0. \end{cases} \tag{4.10}$$

If saturation is active either at the input or at the output or if both signals are constrained, the loop is open and integral action can cause the well-known windup effects. The saturation elements are sector nonlinearities bounded by straight lines passing through the origin with the slopes zero and one.

The characteristic polynomial of the linear loop

$$C_P(s) = N_C(s)N(s) + D_C(s)D(s) \tag{4.11}$$

is supposed to be subdivided according to

$$C_P(s) = \Delta(s)\tilde{D}(s), \tag{4.12}$$

where $\Delta(s)$ is a Hurwitz polynomial of degree n_C and $\tilde{D}(s)$ is a Hurwitz polynomial of degree n. Using observer-based control, $\Delta(s)$ characterizes the dynamics of the state-plus-disturbance observer and $\tilde{D}(s)$ the dynamics of the system controlled by state feedback [1, 2].

4.3 Windup Prevention

It is assumed that disturbance inputs can drive both the input u and the output y beyond the corresponding saturation limits u_0 and y_0. To assure windup prevention, it must first be guaranteed that neither input nor output saturation causes windup effects.

4.3.1 Input Saturation

First, we assume that only the input saturation becomes active. In observer-based compensators, a straightforward method that prevents integral or controller windup is the *observer technique* [1]. During input saturation, the non-Hurwitz polynomial $D_C(s)$ is replaced by the Hurwitz polynomial $\Delta(s)$. This is achieved by using

$$N_u(s) = D_C(s) - \Delta(s) \tag{4.13}$$

for a realization

$$u_C(s) = \frac{N_u(s)}{\Delta(s)} u_s(s) + \frac{N_C(s)}{\Delta(s)} y(s) \tag{4.14}$$

of the compensator. This prevents controller windup (i.e., all windup effects caused by the dynamics of the compensator) and it assures Linear Performance Recovery (LPR). The remaining windup effects are now attributable to the polynomial $\tilde{D}(s)$, i.e., to the dynamics assigned by constant state feedback. After applying the observer technique, the linear part of the loop has the form

$$u(s) = -G_{Lu}(s)u_s(s), \tag{4.15}$$

with

$$G_{Lu}(s) = \frac{\tilde{D}(s) - D(s)}{D(s)}. \tag{4.16}$$

If $G_{Lu}(s)$ does not satisfy the CC there is the danger of plant windup. This can be prevented by an ADE [1].

The dynamics of the controlled plant will be $\tilde{D}(s) = D(s)$ (see Sect. 4.3.2) so that $G_{Lu}(s)$ vanishes. Therefore, the danger of plant windup is no problem here.

4.3.2 Output Saturation

In Chap. 3, a method is described for the prevention of windup caused by saturating sensors. It does not require an additional measure because the disturbance rejecting design is such that the linear loop gain satisfies the CC. The details can be found in Sect. 3.2 but the design equations for the compensating approach shall be repeated here.

Given the characteristic polynomial $C_d(s) = s$ of the considered signal model for step-like disturbances, the design of the compensator goes along the following lines. Start with

$$\tilde{D}(s) = D(s) \quad \text{and} \quad \Delta(s) = \Delta_r(s)N(s), \tag{4.17}$$

where $\Delta_r(s)$ has the form

$$\Delta_r(s) = (s + \alpha)^\kappa, \alpha > 0 \qquad (4.18)$$

and κ is the difference degree of the system. The polynomial $Q_D(s)$ and the constant R_D from the decomposition

$$\Delta_r(s) = Q_D(s)s + R_D \qquad (4.19)$$

finally define the polynomials

$$N_C(s) = R_D D(s), \qquad (4.20)$$

and

$$D_C(s) = s \, Q_D(s)N(s) \qquad (4.21)$$

of the compensator. Increasing values of α in (4.18) improve the attenuation of the disturbance in y while the linear part $G_{Ly}(s)$ in (3.9), namely, (3.10), (3.20) satisfies the CC for arbitrary values $\alpha > 0$.

4.3.3 Input and Output Saturation

The conditions in Sects. 4.3.1 and 4.3.2 assure that windup is prevented when either input or output saturation are active. As already discussed in Sect. 2.3, this does not guarantee stability when saturation takes place both at the input and the output.

The only additional degrees of freedom left for influencing the dynamics of the nonlinear loop are contained in an ADE which, however, can only be realized at the input side. And as already discussed in Sect. 2.3, this can be realized when substituting the polynomial $\Delta(s)$ in the observer technique by another Hurwitz polynomial $\Delta_1(s)$ having the same degree and the same leading coefficient as $\Delta(s)$.

Therefore, the anti-windup scheme in the presence of joint input and output saturation has the structure shown in Fig. 4.1. Here the polynomial $N_{u1}(s)$ has the form

$$N_{u1}(s) = D_C(s) - \Delta_1(s). \qquad (4.22)$$

Interpreting the loop of Fig. 4.1 as a standard structure consisting of a linear part and an isolated nonlinearity, the linear part is a 2×2 system described by

$$\begin{bmatrix} u(s) \\ y(s) \end{bmatrix} = -G^1_{Luy}(s) \begin{bmatrix} u_s(s) \\ y_s(s) \end{bmatrix}. \qquad (4.23)$$

The transfer matrix in (4.23) has the form

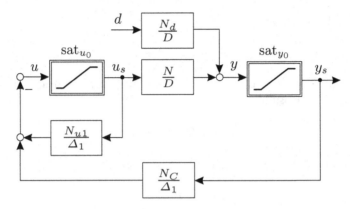

Fig. 4.1 Nonlinear loop with measures to prevent windup due to joint input and output saturation

$$
G_{Luy}^1(s) =
\begin{bmatrix}
\dfrac{N_{u1}(s)}{\Delta_1(s)} & \dfrac{N_C(s)}{\Delta_1(s)} \\[2ex]
-\dfrac{N(s)}{D(s)} & 0
\end{bmatrix}.
\tag{4.24}
$$

Since we are also investigating the structure of Fig. 4.1 for $\Delta_1(s) = \Delta(s)$, the transfer matrix

$$
G_{Luy}(s) =
\begin{bmatrix}
\dfrac{N_u(s)}{\Delta(s)} & \dfrac{N_C(s)}{\Delta(s)} \\[2ex]
-\dfrac{N(s)}{D(s)} & 0
\end{bmatrix}
\tag{4.25}
$$

is of interest too. For the numerical manipulations to test the MIMO CC, it is convenient to use the time-domain representation of (4.23). Given the time-domain representations of the linear systems in Fig. 4.1, namely, of the system

$$
\begin{aligned}
\dot{x}(t) &= A_p x(t) + B_p u_s(t) \\
y(t) &= C_p x(t),
\end{aligned}
\tag{4.26}
$$

and of the compensator

$$
\begin{aligned}
\dot{z}(t) &= A_c z(t) + \begin{bmatrix} B_{cu} & B_{cy} \end{bmatrix} \begin{bmatrix} u_s(t) \\ y_s(t) \end{bmatrix} \\
u(t) &= -C_c z(t) - \begin{bmatrix} 0 & D_{cy} \end{bmatrix} \begin{bmatrix} u_s(t) \\ y_s(t) \end{bmatrix},
\end{aligned}
\tag{4.27}
$$

the time-domain representation of (4.23) has the form

$$\begin{bmatrix} \dot{x}(t) \\ \dot{z}(t) \end{bmatrix} = \begin{bmatrix} A_p & 0 \\ 0 & A_c \end{bmatrix} \begin{bmatrix} x(t) \\ z(t) \end{bmatrix} + \begin{bmatrix} B_p & 0 \\ B_{cu} & B_{cy} \end{bmatrix} \begin{bmatrix} u_s(t) \\ y_s(t) \end{bmatrix}$$

$$\begin{bmatrix} u(t) \\ y(t) \end{bmatrix} = \begin{bmatrix} 0 & -C_c \\ C_p & 0 \end{bmatrix} \begin{bmatrix} x(t) \\ z(t) \end{bmatrix} + \begin{bmatrix} 0 & -D_{cy} \\ 0 & 0 \end{bmatrix} \begin{bmatrix} u_s(t) \\ y_s(t) \end{bmatrix}. \tag{4.28}$$

Comparison of Eqs. (4.28) and (2.24) shows a considerable difference in order. The system for the test of the CC is of the order $4n$ in Chap. 2 whereas here this system is only of the order $2n$, i.e., it coincides with the order of the linear loop. On the one hand, a reduction of dimensions reduces the numerical errors. On the other hand, it is much more intricate to assign the n zeros of $\tilde{D}(s)$ and the κ zeros of $\Delta(s)$ in the compensator design of Chap. 2 than to choose one suitable parameter α here.

In the examples, it will be investigated whether a design parameter α exists such that $G_{Luy}(s)$ (i.e., for $\Delta_1(s) = \Delta(s)$) satisfies the CC and what improvement is possible for a polynomial $\Delta_1(s) \neq \Delta(s)$ "optimized" to allow for an increased α (the "optimization" process is described in Remark 2.1).

Example 4.1 In Example 2.1, a stabilizing control was designed for a nonlinear loop with input and output constraints using a saturation indicator. The transfer functions of the system are

$$G(s) = \frac{N(s)}{D(s)} = \frac{6.25s^2 + 37.5s + 50}{s^4 + 8s^3 + 39s^2 + 62s + 50},$$

and

$$G_d(s) = \frac{N_d(s)}{D(s)} = \frac{20s^3 + 80s^2 + 170s + 300}{s^4 + 8s^3 + 39s^2 + 62s + 50}.$$

In addition to an output saturation with $y_0 = 0.5$, there is also an input saturation with $u_0 = 5$.

For $\Delta_1(s) = \Delta(s)$, the transfer matrix $G_{Luy}(s)$ satisfies the CC for all $0 < \alpha \leq 3.535$. For $\alpha = 3.5$, the compensator has the transfer function

$$G_C(s) = \frac{12.25D(s)}{s(s+7)N(s)}. \tag{4.29}$$

By an "optimization" to find the best pair $(\alpha, \Delta_1(s))$ one obtains

$$\Delta_1(s) = 6.25s^4 + 42.9674s^3 + 156.4597s^2 + 226.4295s + 142.5513. \tag{4.30}$$

With this polynomial the transfer matrix $G_{Luy}^1(s)$ satisfies the CC for all $0.64 < \alpha < 7.403$. For $\alpha = 7.4$, the compensator is characterized by

$$G_C(s) = \frac{54.76D(s)}{s(s+14.8)N(s)}. \tag{4.31}$$

Fig. 4.2 Disturbance
transients of the linear loop

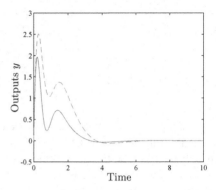

Fig. 4.3 Disturbance
transients of the nonlinear
loop

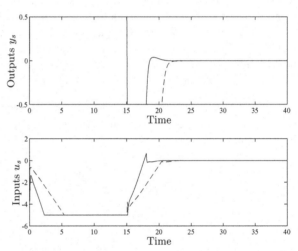

It is obvious that the compensator (4.31) allows a better disturbance attenuation
than the compensator (4.29). Figure 4.2 shows the disturbance step responses in
the unconstrained case with the compensator (4.29) in broken lines and with the
compensator (4.31) in full lines. The improvement is not spectacular. However, the
peak value of the disturbance step response can be reduced by 22% when using the
ADE (i.e., a polynomial $\Delta_1(s) \neq \Delta(s)$).

Applying the same input sequence $d(t) = 1(t) - 1(t - 15)$ as in Example 2.1, the
nonlinear transients shown in Fig. 4.3 result. During the first 15 s the input signal is
not sufficient to reject the disturbance. This was chosen to demonstrate the behavior
of the loop when input and output are constrained simultaneously.

The broken lines show the reaction with the compensator (4.29) and $\Delta_1(s) = \Delta(s)$
and the full lines the reaction with the compensator (4.31) and $\Delta_1(s)$ as in (4.30).

The results in Examples 2.1 and 4.1 are very similar. This is an indication that
the proceeding suggested in Chap. 2 assures indeed stability though it is based on an
assumption without strict proof.

The next two examples demonstrate that it depends on the system whether a stabilizing control can be constructed in the joint presence of input and output saturation or not.

Example 4.2 Consider a very simple system of the order two with one zero. Its transfer behavior is defined by

$$G(s) = \frac{N(s)}{D(s)} = \frac{101s + 101}{s^2 + 2s + 101}.$$

In a loop of Fig. 4.1 with $\Delta_1(s) = \Delta(s)$, there exists no $\alpha > 0$ for which the transfer matrix $G_{Luy}(s)$ satisfies the CC.

When "optimizing" for the best pair $(\alpha, \Delta_1(s))$, one obtains

$$\Delta_1(s) = 101s^2 + 1018s + 509.$$

For this polynomial, the transfer matrix $G_{Luy}^1(s)$ satisfies the CC for all $0 < \alpha \le 0.00429$. For $\alpha = 0.00429$, the compensator is characterized by the polynomials

$$N_C(s) = 0.00429 D(s), \quad D_C = sN(s).$$

This control is far from being satisfactory, because the slowest eigenvalue of the linear loop is located at $s = -0.00429$ so that the disturbance step response only settles after about 1500 s.

Remark 4.1 Given the above results it is obvious that there exist systems where a stabilizing control cannot be parametrized at all. The reason for the bad result in Example 4.2 is the fact that the step response of the system exhibits enormous overshoots. This happens either in the presence of oscillating modes and/or of zeros that are located right of the poles in the s-plane.

Example 4.3 Next, consider a system with real eigenvalues, however, with two zeros right of the poles. Its transfer function is

$$G(s) = \frac{N(s)}{D(s)} = \frac{125(s + 1)^2}{(s + 5)^3}.$$

In the presence of input and output constraints, there is no parameter α for which a compensator designed on the basis of $\tilde{D}(s) = D(s)$ and $\Delta(s) = (s + \alpha)N(s)$ leads to a transfer matrix $G_{Luy}(s)$ satisfying the CC. "Optimizing" for the best pair $(\alpha, \Delta_1(s))$, one obtains the polynomial

$$\Delta_1(s) = 125s^3 + 218.04s^2 + 216.23s + 108.07,$$

for which $G_{Luy}^1(s)$ satisfies the CC for all $0 < \alpha \le 0.0367$. It is obvious that this control is also beyond applicability. The reason here is the enormous overshoot of the system's step response caused by the zeros right of the poles.

The best results can be obtained for well-damped systems without distinct overshoots.

Example 4.4 As another simple example, consider the system with

$$N(s) = 1, \quad D(s) = (s+1)^3, \quad \text{and} \quad N_d(s) = 2s^2 + 3s + 5.$$

The matrix $G_{Luy}(s)$ satisfies the CC for all $0.015 < \alpha \le 1.259$. For $\alpha = 1.25$, the compensator has the transfer function

$$G_C(s) = \frac{1.953125 D(s)}{(s^2 + 3.75s + 4.6875)s}. \tag{4.32}$$

When determining a favorable polynomial $\Delta_1(s)$, namely,

$$\Delta_1(s) = s^3 + 2.7792s^2 + 1.8619s + 0.37986,$$

the matrix $G^1_{Luy}(s)$ satisfies the CC for all $0.23 \le \alpha \le 2.172$. For $\alpha = 2.17$, the compensator has the transfer behavior

$$G_C(s) = \frac{10.218313 D(s)}{(s^2 + 6.51s + 14.1267)s}. \tag{4.33}$$

Comparing the peak amplitudes of the disturbance step responses in the linear loop the second compensator allows an improvement of 35%. To demonstrate the behavior of the nonlinear loop, the following parameters were chosen: $y_0 = 0.5$, $u_0 = 4$, and $d(t) = 1(t) - 1(t - 15)$. Figure 4.4 shows the reactions of the loop with compensator (4.32) and $\Delta_1(s) = \Delta(s)$ in broken lines and with compensator (4.33) and the above $\Delta_1(s)$ in full lines. During the first 15 s the disturbance cannot be rejected.

Fig. 4.4 Disturbance transients of the nonlinear loop

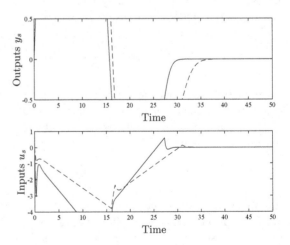

The disturbance step response of the uncontrolled system has a final amplitude 5 whereas the peak amplitude of the disturbance transient in the unconstrained loop with compensator (4.33) is only 1.46.

References

1. P. Hippe, *Windup in Control—Its Effects and Their Prevention* (Springer, Berlin, Heidelberg, New York, London, 2006)
2. P. Hippe, J. Deutscher, *Design of Observer-based Compensators—From the Time to the Frequency Domain* (Springer, Berlin, Heidelberg, New York, London, 2009)

Chapter 5
Windup Prevention in the Joint Presence of Input and Output Saturation—The MIMO Case

5.1 Introductory Remarks

The investigations in Chap. 4 have shown that the stabilization in the presence of input *and* output constraints is not a trivial task. This is exacerbated in the presence of systems with multiple inputs and outputs. Here an additional problem arises, namely, that of *deadlock*. If external disturbances drive both the inputs and the outputs into saturation it can happen that all or some of the input and output signals remain saturated even after the disturbance input has vanished. This property is not due to an inapt compensator design, it is a feature of the uncontrolled system and it depends on its static behavior. There exists a strong tendency for deadlock and a slight tendency for it. In the latter case, an appropriately designed compensator can stabilize the nonlinear loop so that deadlock does not occur. If, however, the system has a strong tendency for deadlock no stabilizing control exists. The reasons for a possible non-existence of a stabilizing control mentioned in Chap. 4 also apply to MIMO systems, of course.

Since the anti-windup design of Chap. 1 is not applicable to MIMO systems the prevention of windup caused by output constraints must use the compensating approach of Chap. 3. The method for anti-windup control in the presence of input constraints is very similar to the one for SISO systems, especially if one uses the frequency-domain design on the basis of coprime Matrix Fraction Descriptions (MFDs). This is also true for the realization of the Additional Dynamic Element (ADE) to improve the performance of the disturbance rejecting control in the input- and output-constrained loop. Also the "optimization" process to find the best solution is the same as in Chap. 4. It is, however, complicated by the increased dimensions of the transfer matrix $G^1_{Luy}(s)$.

As in the SISO case, the stability of the anti-windup control in the joint presence of saturating actuators and sensors is strictly proven.

© The Author(s), under exclusive license to Springer Nature Switzerland AG 2021   61
P. Hippe, *Windup in Control Owing to Sensor Saturation*,
SpringerBriefs in Applied Sciences and Technology,
https://doi.org/10.1007/978-3-030-73133-5_5

5.2 Problem Formulation

We consider asymptotically stable, strictly proper, square, minimum-phase, completely controllable, and observable LTI systems of the order n with input $u_s \in \mathbb{R}^p$ and output $y \in \mathbb{R}^p$. The transfer behavior of these systems is described by

$$y(s) = G(s)u_s(s). \tag{5.1}$$

It is further assumed that constant or step-like changing disturbances $d \in \mathbb{R}^q$ affect the system and that the transfer behavior between these disturbances and y is described by

$$y(s) = G_d(s)d(s). \tag{5.2}$$

The $p \times p$ transfer matrix $G(s)$ is either represented by its right coprime MFD

$$G(s) = N(s)D^{-1}(s), \tag{5.3}$$

or by its left coprime MFD

$$G(s) = \bar{D}^{-1}(s)\bar{N}(s), \tag{5.4}$$

with $D(s)$ column reduced and both $\bar{N}(s)$ and $\bar{D}(s)$ row reduced. The coefficient κ is defined as the maximum difference degree of all outputs, namely,

$$\kappa = \overset{\max}{j} \left\{ \delta_{rj}[\bar{D}(s)] - \delta_{rj}[\bar{N}(s)] \right\}, \tag{5.5}$$

where $\delta_{rj}[\cdot]$ is the jth row degree. The $p \times q$ transfer matrix $G_d(s)$ is represented by its left MFD

$$G_d(s) = \bar{D}^{-1}(s)\bar{N}_d(s), \tag{5.6}$$

with $\bar{D}(s)$ as defined in (5.4). Since $\det N(0) \neq 0$ in minimum-phase systems, constant disturbances can be rejected. The linear compensator of the order n_C is described by

$$u_C(s) = G_C(s)y_s(s), \tag{5.7}$$

where y_s is the output of the sensors. This compensator has been designed as an observer-based compensator for robust rejection of constant disturbances such that the loop with feedback interconnection $u_s = -u_C$ and $y_s = y$ is stable. The transfer matrix $G_C(s)$ is represented by its left coprime MFD

$$G_C(s) = D_C^{-1}(s)N_C(s). \tag{5.8}$$

This compensator contains integral action so that one has

$$D_C(s) = D_C^*(s)D_{Dist}(s),\tag{5.9}$$

with the $p \times p$ diagonal matrix

$$D_{Dist}(s) = \text{diag}\,[s]\,.\tag{5.10}$$

It is supposed that det $D_C^*(s)$ is a Hurwitz polynomial. The design of such compensators is facilitated (in analogy to the SISO case) when $\bar{N}(s)$ is a right divisor of $\Delta(s)$.

The measured output vector $y_s \in \mathbb{R}^p$ is a saturated version of the output y of the system. The components of the nonlinearity $y_s = \text{sat}_{y_0}(y)$ are defined by

$$\text{sat}_{y_0}(y_i) := \begin{cases} y_{0i} & \text{if } y_i > y_{0i} > 0 \\ y_i & \text{if } -y_{0i} \le y_i \le y_{0i}; i = 1, 2, \ldots, p. \\ -y_{0i} & \text{if } y_i < -y_{0i} \end{cases}\tag{5.11}$$

The input $u_s = \text{sat}_{u_0}(u)$ to the system is a saturated version of the command signal

$$u = -u_C,\tag{5.12}$$

where the components of the nonlinearity $u_s = \text{sat}_{u_0}(u)$ are defined by

$$\text{sat}_{u_0}(u_i) := \begin{cases} u_{0i} & \text{if } u_i > u_{0i} > 0 \\ u_i & \text{if } -u_{0i} \le u_i \le u_{0i}; i = 1, 2, \ldots, p. \\ -u_{0i} & \text{if } u_i < -u_{0i} \end{cases}\tag{5.13}$$

If saturation takes place either at the input or at the output or if both signals are constrained, the loop is open and an integral action in the compensator can cause the well-known windup effects. The saturation elements are sector nonlinearities bounded by two straight lines passing the origin with the slopes zero and one.

5.3 Windup Prevention

It is assumed that persistent disturbances can drive both the inputs u_i and the outputs y_i beyond their saturation limits u_{0i} and y_{0i}, $i = 1, 2, \ldots, p$. To assure windup prevention, it must first be guaranteed that neither input nor output saturation causes windup effects.

In what follows, the characteristic polynomial matrix of the linear loop is supposed to be subdivided according to

$$N_C(s)N(s) + D_C(s)D(s) = \Delta(s)\tilde{D}(s),\tag{5.14}$$

where $\Delta(s)$ is a $p \times p$ polynomial matrix with $\Gamma_r[\Delta(s)] = \Gamma_r[D_C(s)]$ and $\tilde{D}(s)$ is a $p \times p$ polynomial matrix with $\Gamma_c[\tilde{D}(s)] = \Gamma_c[D(s)]$. Here $\Gamma_r[\cdot]$ characterizes the highest row-degree coefficient matrix and $\Gamma_c[\cdot]$ the highest column-degree coefficient matrix. Then $\det \Delta(s)$ is a Hurwitz polynomial of degree n_C, characterizing the dynamics of the (state-plus-disturbance) observer and $\det \tilde{D}(s)$ is a Hurwitz polynomial of degree n, characterizing the dynamics of the system controlled by state feedback [1].

5.3.1   Input Saturation

In this subsection, it is assumed that only the input saturation becomes active. As in SISO systems, we use the *observer technique* to prevent integral or controller windup [1, 2]. In the observer technique, one replaces the unstable compensator dynamics ($\det D_C(s)$ contains p zeros at $s = 0$) by the stable observer dynamics ($\det \Delta(s)$ is always a Hurwitz polynomial) in case of input saturation.

Using

$$N_u(s) = D_C(s) - \Delta(s) \tag{5.15}$$

for a realization

$$u_C(s) = \Delta^{-1}(s)N_u(s)u_s(s) + \Delta^{-1}(s)N_C(s)y(s) \tag{5.16}$$

of the compensator, controller windup due to input saturation is prevented and this also assures Linear Performance Recovery (LPR).

If the resulting linear part,

$$u(s) = -G_{Lu}(s)u_s(s), \tag{5.17}$$

of the loop with

$$G_{Lu}(s) = \left[\tilde{D}(s) - D(s)\right]D^{-1}(s) \tag{5.18}$$

violates the Circle Criterion (CC) there is the danger of *plant windup*. The danger of plant windup can be prevented by an ADE [1, 2].

Since the compensating approach is characterized by $\tilde{D}(s) = D(s)$, the transfer matrix $G_{Lu}(s)$ is identically zero so that the danger of plant windup is no problem here.

5.3.2 Output Saturation

A method to prevent windup in MIMO systems subject to sensor saturation, namely, the compensating approach, was presented in Chap. 3. It allows a nearly arbitrarily good attenuation of step-like disturbances, and this in a way that sensor saturation does not cause windup effects, i.e., $G_{Ly}(s)$ in (3.57) satisfies the CC for arbitrary parameters $\alpha > 0$. The details can be found in Sect. 3.4. For convenience, the design equations will be repeated here.

In the compensating approach, the compensator is parametrized by

$$\tilde{D}(s) = D(s), \tag{5.19}$$

and by the $p \times p$ polynomial matrix

$$A(s) = \Delta_r(s)\bar{N}(s), \tag{5.20}$$

where

$$\Delta_r(s) = \text{diag}[\hat{\Delta}_r(s)], \tag{5.21}$$

and $\hat{\Delta}_r(s)$ is a polynomial

$$\hat{\Delta}_r(s) = (s + \alpha)^\kappa, \tag{5.22}$$

with κ the maximum difference degree of all outputs, defined in (5.5).

With the $p \times p$ diagonal polynomial matrix $Q_D(s)$ and the $p \times p$ diagonal matrix R_D from the decomposition

$$\Delta_r(s) = Q_D(s)D_{dis}(s) + R_D, \tag{5.23}$$

the entities in the left MFD (5.8) have the forms

$$D_C(s) = Q_D(s)\bar{N}(s)D_{dis}(s), \tag{5.24}$$

and

$$N_C(s) = R_D\bar{D}(s). \tag{5.25}$$

This compensator assigns the eigenvalues of the loop to the locations defined by the roots of $\det D(s)$, $\det N(s)$, and $\det \Delta_r(s)$ and for arbitrary values $\alpha > 0$ it assures that the open-loop gain

$$G_{Ly}(s) = -N(s)D^{-1}(s)A^{-1}(s)N_C(s) = -\Delta_r^{-1}(s)R_D \tag{5.26}$$

satisfies the CC, because $G_{Ly}(s)$ is a diagonal transfer matrix with diagonal elements

$$G_{Ly}^{ii}(s) = -\frac{R_D^{ii}}{\hat{\Delta}_r(s)} = -\frac{\alpha^{\kappa}}{(s+\alpha)^{\kappa}}, i = 1, 2, \ldots, p \tag{5.27}$$

(see also Sect. 3.4). Increased values of α improve the disturbance attenuation.

5.3.3 Input and Output Saturation

The measures described in Sects. 5.3.1 and 5.3.2 assure that windup problems do not occur when either the inputs u_i or the outputs y_i saturate. Figure 5.1 shows the nonlinear loop with the above-discussed measures.

They guarantee a stable behavior of the nonlinear loop if either the input or the output saturation is active. To obtain a stable behavior of the nonlinear loop also in the joint presence of input and output saturation, the compensator must be designed such that $G_{Luy}(s)$ in

$$\begin{bmatrix} u(s) \\ y(s) \end{bmatrix} = -G_{Luy}(s) \begin{bmatrix} u_s(s) \\ y_s(s) \end{bmatrix}, \tag{5.28}$$

characterizing the linear part of the loop in Fig. 5.1, satisfies the CC. This linear part is now a system with $2p$ inputs and $2p$ outputs. Given the loop in Fig. 5.1 the $2p \times 2p$ transfer matrix $G_{Luy}(s)$ has the form

$$G_{Luy}(s) = \begin{bmatrix} \Delta^{-1}(s)N_u(s) & \Delta^{-1}(s)N_C(s) \\ -N(s)D^{-1}(s) & 0 \end{bmatrix}. \tag{5.29}$$

The above-described compensator is of the order

$$n_C = p\kappa + n_z, \tag{5.30}$$

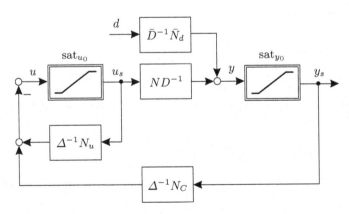

Fig. 5.1 Nonlinear loop with measures to prevent windup due to input saturation or output saturation

where n_z is the number of final zeros of det $N(s)$. Due to the compensating approach the design parameters are reduced to one number α. This does not restrict the performance of the control, but it considerably simplifies the design process of the stabilizing controller described below.

If $G_{Luy}(s)$ violates the CC the danger of an unstable behavior exists as in the SISO case. For systems with $p > 1$ there is the possibility of an additional problem, namely, that of *deadlock*. It turns out that there is a class of MIMO systems, where a compensating control with $\tilde{D}(s) = D(s)$ always leads to the danger of deadlock in the scheme of Fig. 5.1.

In [3], a graphical criterion was presented for the existence of the danger of deadlock. This uses the plots of the frequency responses

$$\text{diag}[\gamma_i]G_{Luy}(j\omega), \quad i = 1, 2, \ldots, 2p \tag{5.31}$$

for all possible combinations of the numbers $0 \le \gamma_i \le 1$. If for $\omega = 0$ one of these plots sits on the real axis left of -1 the danger of deadlock exists. Consequently, it suffices to investigate the behavior of (5.31) for $\omega = 0$.

It is interesting to note that the eigenvalues of $G_{Luy}(0)$ are independent of the polynomial matrix $\Delta(s)$ characterizing the observer dynamics.

Substituting $\tilde{D}(s) = D(s)$ in (5.14) and evaluating the solution of this equation for $s = 0$, one obtains

$$N_C(0) = \Delta(0)D(0)N^{-1}(0), \tag{5.32}$$

because due to the integral action $D_C(0) = 0$. Consequently, (5.15) yields

$$N_u(0) = -\Delta(0). \tag{5.33}$$

Evaluating $G_{Luy}(0)$ with (5.32) and (5.33), one obtains

$$G_{Luy}(0) = \begin{bmatrix} -I & D(0)N^{-1}(0) \\ -N(0)D^{-1}(0) & 0 \end{bmatrix}. \tag{5.34}$$

This demonstrates that in the loop of Fig. 5.1 the danger of deadlock is independent of the observer parameters and of the dynamics of the system. It only depends on the static gain of the system to be controlled.

Thus, the test (5.31) can be simplified. There is no danger of deadlock if for all possible combinations of the numbers $0 \le \gamma_i \le 1$ no eigenvalue of the matrices

$$\text{diag}[\gamma_i]G_{Luy}(0), \quad i = 1, 2, \ldots, 2p \tag{5.35}$$

is located left of -1.

Remark 5.1 The eigenvalue of (5.35) lying farthest to the left will be called the Leftmost Eigenvalue (LME). This LME also defines the intensity of the danger of deadlock (see Remark 5.3).

It turns out that a simple criterion exists indicating the danger of deadlock in a system with input and output constraints.

Deadlock Criterion (DLC) Given a $p \times p$ system with transfer matrix $G(s)$ and $p < 4$. The elements of the $p \times p$ matrix $\Gamma_{SIG}[G(0)]$ are the signs of the corresponding elements of $G(0)$. Then, if and only if

$$\Gamma_{SIG}[G^{-1}(0)] = (\Gamma_{SIG}[G(0)])^T, \tag{5.36}$$

no eigenvalue of (5.35) is located left of -1. Whenever $G(0)$ or $G^{-1}(0)$ contain zero elements, they can be attributed suitable signs. The occurrence of such zeros elements, however, is an indicator that the system is at the limit to the danger of deadlock.

Thus, when the compensator is designed with $\tilde{D}(s) = D(s)$ the nonlinear loop in Fig. 5.1 is not stable if the system does not satisfy the above criterion.

Example 5.1 For demonstration purposes, consider two 2×2 MIMO systems. The transfer behavior of the first is such that

$$G(0) = \begin{bmatrix} 2 & -1 \\ 1 & 0 \end{bmatrix} \quad \text{and} \quad G^{-1}(0) = \begin{bmatrix} 0 & 1 \\ -1 & 2 \end{bmatrix},$$

leading to

$$(\Gamma_{SIG}[G(0)]) = \begin{bmatrix} + & - \\ + & + \end{bmatrix},$$

and

$$\Gamma_{SIG}[G^{-1}(0)] = \begin{bmatrix} + & + \\ - & + \end{bmatrix}.$$

Thus, condition (5.36) is satisfied and there is no danger of deadlock. The system is, however, at the limit to this danger.

The transfer behavior of the second system is such that

$$G(0) = \begin{bmatrix} 2 & 1 \\ 1 & 1 \end{bmatrix} \quad \text{and} \quad G^{-1}(0) = \begin{bmatrix} 1 & -1 \\ -1 & 2 \end{bmatrix},$$

so that

$$(\Gamma_{SIG}[G(0)]) = \begin{bmatrix} + & + \\ + & + \end{bmatrix},$$

and

$$\Gamma_{SIG}[G^{-1}(0)] = \begin{bmatrix} + & - \\ - & + \end{bmatrix}.$$

Here condition (5.36) is not satisfied so that for arbitrary dynamics of the system and of the observer deadlock can occur in the loop of Fig. 5.1 if the input and output constraints become active simultaneously.

As discussed in the SISO case (see Chap. 4), an increased overshoot behavior of the system can lead to problems in the design of a stabilizing compensator. This is, of course, also true for MIMO systems. The tendency for deadlock is another reason why a stabilizing compensator may not exist. This will be discussed in more detail later.

Remark 5.2 Unfortunately, the DLC is only sufficient for systems with $p < 4$ inputs. For systems with $p \geq 4$, it can happen that the DLC is satisfied but there exist eigenvalues of (5.31) sitting on the real axis left of -1 for $\omega = 0$. Though this seems to reduce the importance of the DLC, it should be kept in mind that at the one hand, the number of practical applications for systems with dimensions $p \geq 4$ will not be very numerous. On the other hand, for $p >> 3$ it is all but easy to find a system with all elements of $G_{ij}(0) \neq 0$ such that the DLC is satisfied. If it is, the tendency of deadlock (see discussions below) is not very intensive so that the control discussed in Sect. 5.5 can possibly cope with it.

5.4 Windup Prevention When both Input and Output Saturation Are Activated

As already discussed in Sects. 2.3 and 4.3.3, the only degrees of freedom left for a modification of the loop gain are contained in an ADE at the input side of the loop. Its influence can be taken into account in the SISO case by replacing the observer polynomial $\Delta(s)$ by some other Hurwitz polynomial $\Delta_1(s)$ having the same degree and the same leading coefficient as $\Delta(s)$.

Here in the MIMO case this corresponds to the choice of a $p \times p$ polynomial matrix $\Delta_1(s)$ such that $\det \Delta_1(s)$ is Hurwitz and

$$\Gamma_r[\Delta_1(s)] = \Gamma_r[\Delta(s)], \tag{5.37}$$

and

$$\delta_{ri}[\Delta_1(s)] = \delta_{ri}[\Delta(s)], i = 1, 2, \ldots, p. \tag{5.38}$$

This yields additional degrees of freedom for the design of a stabilizing control.

Figure 5.2 shows the scheme for the prevention of windup in the joint presence of input and output saturation.

To obtain LPR also here the polynomial matrix $N_{u1}(s)$ must have the form

$$N_{u1}(s) = D_C(s) - \Delta_1(s), \tag{5.39}$$

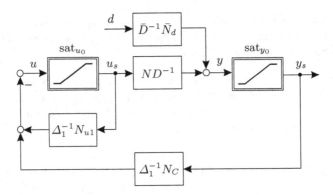

Fig. 5.2 Nonlinear loop with measures to prevent windup due to joint input and output saturation

where (5.37) and (5.38) assure the absence of algebraic loops in the scheme of Fig. 5.2.

Now the transfer behavior (5.28) of the linear part of the loop in Fig. 5.2 is characterized by the $2p \times 2p$ transfer matrix

$$G^1_{Luy}(s) = \begin{bmatrix} \Delta_1^{-1}(s)N_{u1}(s) & \Delta_1^{-1}(s)N_C(s) \\ -N(s)D^{-1}(s) & 0 \end{bmatrix}. \tag{5.40}$$

If the design parameters in $\Delta_r(s)$ and $\Delta_1(s)$ are chosen such that $G^1_{Luy}(s)$ satisfies the CC, the nonlinear loop in Fig. 5.2 exhibits a stable behavior. The polynomial matrix $\Delta_r(s)$ is chosen as a diagonal matrix with identical entries whereas in $\Delta_1(s)$ all degrees of freedom should be exploited.

For the test of the CC, it is again convenient to use the time-domain representation of (5.28) both for $\Delta_1(s) = \Delta(s)$ and $\Delta_1(s) \neq \Delta(s)$. This has the same form as presented in Sect. 4.3 (see Eqs. (4.26)–(4.28)).

The matrix characterizing the danger of deadlock in the joint presence of input and output saturation now has the form

$$G^1_{Luy}(0) = \begin{bmatrix} -I & \Delta_1^{-1}(0)\Delta(0)D(0)N^{-1}(0) \\ -N(0)D^{-1}(0) & 0 \end{bmatrix}, \tag{5.41}$$

which shows that by $\Delta_1(s) \neq \Delta(s)$ the danger of deadlock can be influenced. If in systems with $p < 4$ the danger of deadlock is removed by $\Delta_1(s)$, one has

$$\Gamma_{SIG}[\Delta_1^{-1}(0)\Delta(0)D(0)N^{-1}(0)] = (\Gamma_{SIG}[G(0)])^T. \tag{5.42}$$

Remark 5.3 The system is said to have a slight tendency for deadlock if the LME of (5.35) is in the vicinity of -1, i.e., if it is located between -1 and -2. The system is said to have a strong tendency for deadlock if the LME lies left of -2.

If a system has a slight tendency for deadlock it is often possible to find a stabilizing compensator, i.e., a pair $(\alpha, \Delta_1(s))$ such that $G^1_{Luy}(s)$ satisfies the CC. But the corresponding α can already be comparatively small. If, however, the system has a strong tendency for deadlock, one can only find a pair $(\alpha, \Delta_1(s))$ with a very small $\alpha > 0$ or no such pair at all. Therefore, in general, the problem of joint input and output constraints only has an acceptable solution when the DLC is satisfied.

This shall be demonstrated by the following three examples.

Example 5.2 First consider a 2×2 system described by its MFDs $N(s)D^{-1}(s)$, $\bar{D}^{-1}(s)\bar{N}(s)$, and $\bar{D}^{-1}(s)\bar{N}_d(s)$ with

$$N(s) = \begin{bmatrix} 3 & 1 \\ 1 & 3 \end{bmatrix}, \quad D(s) = \begin{bmatrix} s^2 + 3s + 2.25 & 1 \\ -s - 1 & s + 1 \end{bmatrix},$$

$$\bar{N}(s) = \begin{bmatrix} 4 & 2 \\ 0 & 2 \end{bmatrix},$$

$$\bar{D}(s) = \begin{bmatrix} 1.5s^2 + 3.5s + 1.875 & -0.5s^2 - 0.5s + 1.375 \\ -s - 1 & s + 1 \end{bmatrix},$$

and

$$\bar{N}_d(s) = \begin{bmatrix} -3 & -4 \\ 0.5 & 1 \end{bmatrix}.$$

Since

$$G(0) = N(0)D^{-1}(0) = \begin{bmatrix} 4 & -0.75 \\ 4 & 5.75 \end{bmatrix} \frac{1}{3.25},$$

and

$$G^{-1}(0) = \begin{bmatrix} 0.71875 & 0.09375 \\ -0.5 & 0.5 \end{bmatrix},$$

one has

$$(\Gamma_{SIG}[G(0)]) = \begin{bmatrix} + & - \\ + & + \end{bmatrix},$$

and

$$\Gamma_{SIG}[G^{-1}(0)] = \begin{bmatrix} + & + \\ - & + \end{bmatrix},$$

so that condition (5.36) of the DLC is satisfied. Consequently, with a compensator designed on the basis of $\tilde{D}(s) = D(s)$ and $\Delta_1(s) = \Delta(s)$ there is no danger of deadlock. And since $G(0)$ and $G^{-1}(0)$ contain no zero elements one is not at the limit to the danger of deadlock. Since $\kappa = 2$ the parametrizing matrix in $\Delta(s) = \Delta_r(s)\tilde{N}(s)$ has the form

$$\Delta_r(s) = \begin{bmatrix} (s+\alpha)^2 & 0 \\ 0 & (s+\alpha)^2 \end{bmatrix}.$$

For $\Delta_1(s) = \Delta(s)$ (i.e., in the loop of Fig. 5.1), the CC is satisfied for all $0.009 \leq \alpha \leq 1.54$.

Improved results can be obtained in the loop of Fig. 5.2 when looking for the best pair $(\alpha, \Delta_1(s))$, i.e., for a pair with the biggest α. The search process has already been described in Remark 2.1. With

$$\Delta_1(s) = \begin{bmatrix} 4s^2 + 10.34151s + 5.485425 \\ -3.401163s - 1.842269 \end{bmatrix}$$

$$\begin{matrix} 2s^2 - 1.403859s + 3.401315 \\ 2s^2 + 18.98671s + 12.58242 \end{matrix} \Bigg],$$

one obtains a stabilizing control for all $0.43 \leq \alpha \leq 2.57$. Due to the increased value of α this gives a better disturbance rejection. Choosing the biggest α, namely, $\alpha = 2.57$ the compensator is characterized by the polynomial matrices

$$N_C(s) = \begin{bmatrix} 6.6049 & 0 \\ 0 & 6.6049 \end{bmatrix} \bar{D}(s),$$

$$D_C(s) = \begin{bmatrix} s^2 + 5.14s & 0 \\ 0 & s^2 + 5.14s \end{bmatrix} \bar{N}(s),$$

and

$$N_{u1}(s) = \begin{bmatrix} 10.21849s - 5.485425 & 11.683859s - 3.401315 \\ 3.401163s + 1.842269 & -8.70671s - 12.58242 \end{bmatrix}.$$

5.5 How to Reproduce Deadlock in Simulations

If a MIMO system exhibits the danger of deadlock, it can be of interest to reproduce the deadlock in simulations. Deadlock does not occur for all values of the input and output saturation limits. Especially, if there is only a slight tendency for deadlock, it usually only occurs for very specific relative values of the input and output saturation limits. For 2×2 systems, a trial-and-error search for these values may be practicable, but even here, the search for a set of appropriate four values may be a cumbersome task. Therefore, it would be helpful to have a method at hand to find the relative values of the u_{0i} and y_{0i}, $i = 1, 2, \ldots, p$ in a systematic way.

Given the static gain matrix (5.34), the entries in diag$[\gamma_i]$ (see (5.35)) can be represented by a row vector

$$ga = \begin{bmatrix} \gamma_1, \ldots, \gamma_p, & \gamma_{p+1}, \ldots, \gamma_{2p} \end{bmatrix}, \qquad (5.43)$$

where the first p entries represent the (linearized) gains of the saturation elements for the inputs $u_1, u_2, \ldots, u_p$ and the second p entries represent the (linearized) gains of the saturation elements for the outputs $y_1, y_2, \ldots, y_p$. For given amplitudes of the signals, the linearized gains decrease for decreasing values u_{0i} or y_{0i}, respectively.

To find the LME with the aid of (5.35), one does not have to compute the eigenvalues for all possible combinations of the γ_i, $i = 1, 2, \ldots, 2p$ within the limits $0 \le \gamma_i \le 1$ (a task of infinite combinations). Since the extreme locations of the eigenvalues of (5.35) occur for the extreme values of the γ_i it is sufficient to compute the eigenvalues for all possible combinations of the γ_i, $i = 1, 2, \ldots, 2p$ having the values 0 and 1.

Plotting the pairs (eigenvalue, ga) for all combinations of the γ_i, $i = 1, 2, \ldots, 2p$ is still a paper-consuming process. Therefore, one should only print the combinations giving rise to eigenvalues left of -1. To obtain the LME only, one can shift this print limit further left until the output is only the pair (LME, ga). This restriction to the LME makes sense because the reproduction of a deadlock is easiest when the corresponding eigenvalue of (5.35) is as far left of -1 as possible.

For a 3×3 system, the outcome will typically have the form

$$
\begin{aligned}
\text{LME} &= -2.01 \quad ; \quad ga_1 = [1\,0\,1 \quad 0\,1\,1], \\
\text{LME} &= -2.01 \quad ; \quad ga_2 = [0\,1\,0 \quad 1\,0\,0].
\end{aligned}
\qquad (5.44)
$$

The vector ga_1 contains the following information. The linearized gains for the saturation elements of $[u_1, u_2, u_3]$ have the values $[1\,0\,1]$ and the linearized gains for the saturation elements of $[y_1, y_2, y_3]$ have the values $[0\,1\,1]$.

This can be interpreted in the following way. If temporary external disturbances activate all six saturation elements such that u_1 and u_3 are not much restricted while u_2 is exceeding its saturation limit considerably and if, at the same time, y_2 and y_3 are not much restricted while y_1 exceeds its saturation limit considerably, there is the danger that after the disturbances have decayed the loop remains in a deadlock situation.

To reproduce this in simulations, first, select saturation limits $u_{0i} = u_{0is}$ and $y_{0i} = y_{0is}$, $i = 1, 2, \ldots, p$ such that the chosen disturbance inputs cause all inputs and outputs to saturate simultaneously for a certain period of time. Using the information contained in ga_1 one should then assign the saturation limits as $u_{01} = u_{01s}$, $u_{02} = 0.1u_{02s}$, $u_{03} = u_{03s}$ and $y_{01} = 0.1y_{01s}$, $y_{02} = y_{02s}$, $y_{03} = y_{03s}$.

This first attempt may not lead to the desired deadlock. The obvious second attempt is to use the relative gains suggested by the vector ga_2. In the examples, this always worked. But even if this would not be successful, the actual saturation limits causing deadlock will be nearby.

Though the above-described method seems somehow complicated it is decidedly simpler than the search for a point $\{u_{01}, u_{02}, u_{03}, y_{01}, y_{02}, y_{03}\}$ in a six-dimensional space by trial and error.

The examples will shed more light onto this.

5.6   Illustrative Examples

In this section, examples for systems with 2, 3, and 4 inputs and outputs are investigated. The main focus is on how the intensity of deadlock affects the achievable dynamics. The best results are always obtained for systems with no tendency for deadlock.

The next system has the same denominator matrix as in Example 5.2. Its modified numerator matrix, however, leads to a slight tendency for deadlock.

Example 5.3 Now the numerator $N(s)$ of the system of Example 5.2 is slightly modified to

$$N(s) = \begin{bmatrix} 2 & 1.2 \\ 1 & 4 \end{bmatrix},$$

while $D(s)$ remains the same. The left coprime MFD of the system is now characterized by

$$\bar{D}(s) = \begin{bmatrix} 3.4s + 3.4 & -2.176s - 2.176 \\ 10s^2 + 15.3125s + 5.3125 & -3s^2 + 0.4s + 7.65 \end{bmatrix} \frac{1}{3.4},$$

$$\bar{N}(s) = \begin{bmatrix} 0 & -1.36 \\ 5 & 5.875 \end{bmatrix},$$

and $\bar{N}_d(s)$ is the same as in Example 5.2.

Since

$$G(0) = N(0)D^{-1}(0) = \begin{bmatrix} 0.64 & 0.14 \\ 1 & 1.6 \end{bmatrix} \frac{1}{0.65},$$

and

$$G^{-1}(0) = \begin{bmatrix} 1 & -0.0875 \\ -0.625 & 0.4 \end{bmatrix} \frac{1}{0.85},$$

one has

$$(\Gamma_{SIG}[G(0)]) = \begin{bmatrix} + & + \\ + & + \end{bmatrix},$$

and

$$\Gamma_{SIG}[G^{-1}(0)] = \begin{bmatrix} + & - \\ - & + \end{bmatrix},$$

so that the condition (5.36) of the DLC is not satisfied and consequently a compensator designed on the basis of $\tilde{D}(s) = D(s)$ and $\Delta_1(s) = \Delta(s)$ will give rise to a slight danger of deadlock, because the LME is located at $s = -1.139$ (see also Remark 5.3). The corresponding vectors ga_1 and ga_2 (see (5.43)) have the forms $ga_1 = [1\ 0\ \ 0\ 1]$ and $ga_2 = [0\ 1\ \ 1\ 0]$.

Following the approach of Sect. 5.5, first select the limits u_{0is} and y_{0is}. For disturbance inputs $d_1(t) = 1(t) - 1(t - 10)$, $d_2(t) = 1(t) - 1(t - 10)$, one obtains $u_{01s} = u_{02s} = 2$ and $y_{01s} = 1$, $y_{02s} = 2$.

For a simulation with $\Delta_1(s) = \Delta(s)$, we choose

$$\Delta_r(s) = \text{diag}[(s + 1.9)^2],$$

so that the compensator is characterized by

$$D_C(s) = \begin{bmatrix} s^2 + 3.8s & 0 \\ 0 & s^2 + 3.8s \end{bmatrix} \bar{N}(s),$$

and

$$N_C(s) = \begin{bmatrix} 3.61 & 0 \\ 0 & 3.61 \end{bmatrix} \bar{D}(s).$$

Starting with the vector ga_1, the saturation limits are $u_{01} = 2$, $u_{02} = 0.2$ and $y_{01} = 0.1$, $y_{02} = 2$. This leads to a deadlock with $u_1(\infty) = 2$, $u_2(\infty) = -0.2$, $y_1(\infty) = 0.1$, $y_2(\infty) = 2$. In view of the vector ga_2, one can use $u_{01} = 0.2$, $u_{02} = 2$ and $y_{01} = 1$, $y_{02} = 0.2$. This leads to a deadlock with $u_1(\infty) = 0.2$, $u_2(\infty) = -2$, $y_1(\infty) = -0.2338$, $y_2(\infty) = -0.2$. It is interesting to note that in this deadlock situation only u_1, u_2, and y_2 stay at their saturation limits while y_1 is somewhere in its linear range.

When trying to find the best pair $(\alpha, \Delta_1(s))$, one obtains

$$\Delta_1(s) = \begin{bmatrix} 4.989201s + 3.17947 & -1.36s^2 - 9.205802s - 6.826278 \\ 5s^2 + 9.346363s + 1.911647 & 5.875s^2 + 9.608196s + 11.19694 \end{bmatrix}.$$

With this polynomial matrix a compensator parametrized by

$$\Delta_r(s) = \begin{bmatrix} (s+\alpha)^2 & 0 \\ 0 & (s+\alpha)^2 \end{bmatrix}$$

assures a stable behavior of the nonlinear loop for all $0.48 \leq \alpha \leq 1.958$, i.e., with these parameters the matrix $G_{Luy}^1(s)$ satisfies the CC. For $\alpha = 1.9$, the compensator has the above transfer behavior, but now its realization is according to Fig. 5.2 with

$$N_{u1}(s) = \begin{bmatrix} -4.989201s - 3.17947 & 4.037802s + 6.826278 \\ 9.653637s - 1.911647 & 12.716804s - 11.19694 \end{bmatrix}.$$

With

$$\Delta_1^{-1}(0)\Delta(0)D(0)N^{-1}(0) = \begin{bmatrix} 1.62232 & 0.607958 \\ 0.226787 & 0.621625 \end{bmatrix}$$

condition (5.42) is now also satisfied.

Therefore, neither the danger of deadlock nor of windup exists.

Compared to Example 5.2, the value of α is now smaller so that disturbance rejection is not as good as in Example 5.2. The compensator can nevertheless be considered satisfactory. This is a consequence of the fact that the tendency for deadlock is not very intensive.

The next example demonstrates that an increased tendency for deadlock makes it difficult to find a stabilizing control.

Example 5.4 Consider again the 2×2 system studied in Example 5.2. Now the numerator of the system is characterized by

$$N(s) = \begin{bmatrix} 1 & 2 \\ 1 & 3 \end{bmatrix},$$

with $D(s)$ is as above, and the left coprime MFD of the system is characterized by

$$\bar{N}(s) = \begin{bmatrix} 12 & 17 \\ 0 & 0.75 \end{bmatrix},$$

$$\bar{D}(s) = \begin{bmatrix} 36s^2 + 40s + 1 & -24s^2 - 21s + 9 \\ -3s - 3 & 2.25s + 2.25 \end{bmatrix},$$

and $\bar{N}_d(s)$ as in Example 5.2. Since

$$G(0) = N(0)D^{-1}(0) = \begin{bmatrix} 3 & 3.5 \\ 4 & 5.75 \end{bmatrix} \frac{1}{3.25},$$

and

$$G^{-1}(0) = \begin{bmatrix} 5.75 & -3.5 \\ -4 & 3 \end{bmatrix},$$

one has

$$(\Gamma_{SIG}[G(0)]) = \begin{bmatrix} + & + \\ + & + \end{bmatrix},$$

and

$$\Gamma_{SIG}[G^{-1}(0)] = \begin{bmatrix} + & - \\ - & + \end{bmatrix},$$

so that condition (5.36) of the DLC is not satisfied and a compensator designed on the basis of $\tilde{D}(s) = D(s)$ and $\Delta_1(s) = \Delta(s)$ will give rise to the danger of deadlock.

The LME is located at $s = -2.635$ and the corresponding vectors ga_i have the forms $ga_1 = [1\ 0\ \ 0\ 1]$ and $ga_2 = [0\ 1\ \ 1\ 0]$. To reproduce deadlock in simulation, we follow the procedure of Sect. 5.5. With the disturbance inputs $d_1(t) = 1(t) - 1(t - 10)$, $d_2(t) = -1(t) + 1(t - 10)$ the saturation limits u_{0is} and y_{0is} are $u_{01s} = 0.8$, $u_{02s} = 0.5$, and $y_{01s} = y_{02s} = 0.15$.

With $\Delta_1(s) = \Delta(s)$ and

$$\Delta_r(s) = \mathrm{diag}[(s + 1)^2],$$

one obtains a compensator characterized by

$$N_C(s) = \bar{D}(s)$$

and

$$D_C(s) = \begin{bmatrix} s^2 + 2s & 0 \\ 0 & s^2 + 2s \end{bmatrix} \bar{N}(s).$$

The vector ga_1 suggests the saturation limits $u_{01} = 0.8$, $u_{02} = 0.05$ and $y_{01} = 0.015$, $y_{02} = 0.15$. With these saturation limits deadlock occurs. The signals remain at $u_1(\infty) = -0.8$, $u_2(\infty) = 0.05$, $y_1(\infty) = -0.015$, and $y_2(\infty) = -0.15$. Deadlock also occurs when using the saturation limits suggested by the vector ga_2, namely, $u_{01} = 0.08$, $u_{02} = 0.5$ and $y_{01} = 0.15$, $y_{02} = 0.015$. Then, the signals remain at $u_1(\infty) = 0.08$, $u_2(\infty) = -0.5$, $y_1(\infty) = -0.15$, and $y_2(\infty) = -0.015$.

After tedious "optimizations," one can find a stabilizing compensator for this system. With

$$\Delta_1(s) = \begin{bmatrix} 12s^2 + 0.162847s - 0.0739327 & 17s^2 + 0.791944s + 0.0617729 \\ -0.650022s - 0.0322183 & 0.75s^2 + 0.539912s + 0.0246579 \end{bmatrix}$$

and

$$\Delta_r(s) = \begin{bmatrix} (s+0.025)^2 & 0 \\ 0 & (s+0.025)^2 \end{bmatrix},$$

one obtains a stabilizing control. The compensator is now characterized by

$$N_C(s) = \begin{bmatrix} 0.000625 & 0 \\ 0 & 0.000625 \end{bmatrix} \bar{D}(s),$$

and

$$D_C(s) = \begin{bmatrix} s^2+0.05s & 0 \\ 0 & s^2+0.05s \end{bmatrix} \bar{N}(s),$$

and with

$$N_{u1}(s) = \begin{bmatrix} 0.437153s+0.0739327 & 0.058056s-0.0617729 \\ 0.650022s+0.0322183 & -0.502412s-0.0246579 \end{bmatrix}$$

it can be realized according to Fig. 5.2. With

$$\Delta_1^{-1}(0)\Delta(0)D(0)N^{-1}(0) = \begin{bmatrix} 0.784935 & 0.310017 \\ 0.949564 & 0.462102 \end{bmatrix}$$

condition (5.42) is now also satisfied. Though this compensator assures a stable behavior it is not satisfactory because the disturbance step responses of the system settle after about 5 s whereas the disturbance transients of the loop of Fig. 5.2 do so only after about 350 s. This is a consequence of the fact that the system has a strong tendency for deadlock.

A curiosity about this example is the fact that with the above $\Delta_1(s)$ the CC is not satisfied for any other value of α.

The next example demonstrates that yet another increase of the tendency for deadlock prevents the design of a stabilizing compensator.

Example 5.5 The system considered now has the numerator matrix

$$N(s) = \begin{bmatrix} 1 & 2 \\ 1 & 2.6 \end{bmatrix},$$

and $D(s)$ is as above. The left coprime MFD of this system is characterized by

$$\bar{N}(s) = \begin{bmatrix} 0 & -0.5 \\ 10.8 & 15 \end{bmatrix},$$

and

$$\bar{D}(s) = \left[\begin{array}{cc} 3(s+1) & -2.5(s+1) \\ 46.8s^2 + 50.4s - 2.7 & -36s^2 - 33s + 12 \end{array} \right].$$

With

$$G(0) = N(0)D^{-1}(0) = \left[\begin{array}{cc} 3 & 3.5 \\ 3.6 & 4.85 \end{array} \right] \frac{1}{3.25},$$

and

$$G^{-1}(0) = \left[\begin{array}{cc} 24.25 & -17.5 \\ -18 & 15 \end{array} \right] \frac{1}{3},$$

the DLC is not satisfied and the LME is located at -3.09. For this system, no stabilizing $\Delta_1(s)$ can be obtained.

Remark 5.4 The above results demonstrate that for systems with a strong tendency for deadlock no satisfactory control or no stabilizing control at all can be found. The tendency for slow achievable dynamics is directly related to the intensity of the danger of deadlock. It is therefore not advisable to design anti-windup control for systems with input and output constraints when the tendency for deadlock exists.

The next examples demonstrate that an overshoot behavior (either due to badly damped eigenvalues and/or to zeros right of the poles) also restricts the achievable dynamics of the stabilizing control as in SISO systems.

Example 5.6 We now consider a 2×2 system with badly damped poles at $-1 \pm 3j$ and $-2 \pm 6j$ and with two zeros at $s = -2$, characterized by the MFDs $G(s) = N(s)D^{-1}(s) = \bar{D}^{-1}(s)\bar{N}(s)$ and $G_d(s) = \bar{D}^{-1}(s)\bar{N}_d(s)$ with

$$N(s) = \left[\begin{array}{cc} s+1 & 1 \\ -1 & s+3 \end{array} \right],$$

$$D(s) = \left[\begin{array}{cc} s^2 + 2s + 10 & 3s + 5 \\ 0 & s^2 + 4s + 40 \end{array} \right],$$

$$\bar{D}(s) = \left[\begin{array}{cc} 11s^2 + 24s + 136.25 & 24.25s + 6.25 \\ -25s - 5 & 11s^2 + 42s + 355 \end{array} \right] \frac{1}{11},$$

$$\bar{N}(s) = \left[\begin{array}{cc} 11s + 13 & 2.25 \\ -36 & 11s + 31 \end{array} \right] \frac{1}{11},$$

and

$$\bar{N}_d(s) = \begin{bmatrix} 25s + 14 & 20s + 12 \\ 16s + 90 & 14s + 80 \end{bmatrix}.$$

The maximum difference degree of all outputs is $\kappa = 1$ so that the design parameter of the control is

$$\Delta_r(s) = \begin{bmatrix} s + \alpha & 0 \\ 0 & s + \alpha \end{bmatrix}.$$

With

$$G(0) = \begin{bmatrix} 0.1 & 0.0125 \\ -0.1 & 0.0875 \end{bmatrix},$$

and

$$G^{-1}(0) = \begin{bmatrix} 8.75 & -1.25 \\ 10 & 10 \end{bmatrix},$$

there is no danger of deadlock.

The system cannot be stabilized with $\Delta_1(s) = \Delta(s)$. Searching for the best pair $(\alpha, \Delta_1(s))$ one obtains

$$\Delta_1(s) = \begin{bmatrix} s^2 + 3.25248s + 4.56911 & 0.668816s + 0.905461 \\ -2.78564s - 4.15266 & s^2 + 3.12434s + 4.89030 \end{bmatrix}.$$

With this polynomial matrix the transfer matrix $G^1_{Luy}(s)$ satisfies the CC for all $0 < \alpha \le 0.3$. Using the fastest control, i.e., $\alpha = 0.3$, the compensator is characterized by the polynomial matrices

$$N_C(s) = \begin{bmatrix} 0.3 & 0 \\ 0 & 0.3 \end{bmatrix} \bar{D}(s),$$

and

$$D_C(s) = \begin{bmatrix} s & 0 \\ 0 & s \end{bmatrix} \bar{N}(s).$$

The badly damped eigenvalues of this system do not prevent the existence of a stabilizing compensator but the results are definitely degraded. This will be demonstrated by the following example.

Example 5.7 We now consider a 2×2 system with real poles at $-8, -5, -5$, and -2 and two zeros at $s = -2$, characterized by $N(s)$ as in Example 5.6 and

$$D(s) = \begin{bmatrix} s^2 + 7s + 10 & 3s + 5 \\ 0 & s^2 + 13s + 40 \end{bmatrix}.$$

The matrices of the left coprime MFD now have the form

$$\bar{D}(s) = \begin{bmatrix} 117s^2 + 954s + 2215 & 235s - 305 \\ -207s - 675 & 117s^2 + 1386s + 2565 \end{bmatrix} \frac{1}{117},$$

and

$$\bar{N}(s) = \begin{bmatrix} 117s + 252 & 1 \\ -324 & 117s + 216 \end{bmatrix} \frac{1}{117},$$

and $\bar{N}_d(s)$ is the same as in Example 5.6. The maximum difference degree of all outputs is again $\kappa = 1$ so that

$$\Delta_r(s) = \begin{bmatrix} s + \alpha & 0 \\ 0 & s + \alpha \end{bmatrix}$$

parametrizes the compensator. The static behavior is the same as in Example 5.6. So there is no danger of deadlock.

This system can be stabilized with $\Delta_1(s) = \Delta(s)$ for all $0.05 \le \alpha \le 1.94$.

When looking for the best pair $(\alpha, \Delta_1(s))$ one obtains

$$\Delta_1(s) = \begin{bmatrix} s^2 + 4.59162s + 6.86124 & 2.51125s + 1.12861 \\ -4.04236s - 6.12187 & s^2 + 7.74491s + 10.7039 \end{bmatrix}.$$

With this polynomial matrix the transfer matrix $G^1_{Luy}(s)$ satisfies the CC for all $0 < \alpha \le 4.1$. Using the fastest control, i.e., $\alpha = 4.1$, the compensator is characterized by its left coprime MFDs

$$N_C(s) = \begin{bmatrix} 4.1 & 0 \\ 0 & 4.1 \end{bmatrix} \bar{D}(s),$$

and

$$D_C(s) = \begin{bmatrix} s & 0 \\ 0 & s \end{bmatrix} \bar{N}(s).$$

Though the system's basic speed and structure are about the same as in Example 5.6, the resulting control is much faster and the system can even be stabilized in the control structure of Fig. 5.1. The disturbance step responses of the loop in Example

5.6 die out after about 25 s whereas the reaction to the same input signals settles already after about 4 s here.

In the sequel, two MIMO systems with three inputs and outputs are investigated.

Example 5.8 Given a system with three inputs and outputs with real poles at $-1, -1, -1, -2, -2, -2$ and zeros at $-2 \pm 2j, -4$, characterized by its MFDs $G(s) = N(s)D^{-1}(s) = \bar{D}^{-1}(s)\bar{N}(s)$ and $G_d(s) = \bar{D}^{-1}(s)\bar{N}_d(s)$ with

$$N(s) = \begin{bmatrix} s+4 & -1 & -1 \\ 0 & s+4 & 0 \\ 8 & -4 & s \end{bmatrix},$$

$$D(s) = \begin{bmatrix} s^2+2s+1 & -s-1 & 0 \\ s+1 & s^2+4s+3 & 0 \\ 2s+2 & -4s-4 & s^2+3s+2 \end{bmatrix},$$

$$\bar{D}(s) = \begin{bmatrix} s^3+2s^2-s-2 & -9s-9 & 0 \\ s^2+3s+2 & s^2+6s+5 & 0 \\ -2s-2 & 0 & s+1 \end{bmatrix},$$

$$\bar{N}(s) = \begin{bmatrix} s^2+4s & -10 & -s+1 \\ s+4 & s+6 & -1 \\ -2 & 0 & 1 \end{bmatrix},$$

and

$$\bar{N}_d(s) = \begin{bmatrix} 2s+4 & -3 & 5 \\ -3 & 4s-10 & 8 \\ 5 & -6 & 10 \end{bmatrix}.$$

The maximum difference degree κ of all outputs is $\kappa = 1$ so that the parametrizing matrix has the form

$$\Delta_r(s) = \begin{bmatrix} s+\alpha & 0 & 0 \\ 0 & s+\alpha & 0 \\ 0 & 0 & s+\alpha \end{bmatrix}.$$

With

$$G(0) = \begin{bmatrix} 4.5 & 0.5 & -0.5 \\ -1 & 1 & 0 \\ 7 & 1 & 0 \end{bmatrix},$$

and

$$G^{-1}(0) = \begin{bmatrix} 0 & -0.125 & 0.125 \\ 0 & 0.875 & 0.125 \\ -2 & -0.25 & 1.25 \end{bmatrix},$$

one has

$$\Gamma_{SIG}[G(0)] = \begin{bmatrix} + & + & - \\ - & + & - \\ + & + & + \end{bmatrix},$$

and

$$\Gamma_{SIG}[G^{-1}(0)] = \begin{bmatrix} + & - & + \\ + & + & + \\ - & - & + \end{bmatrix},$$

so that no danger of deadlock exists. The system is, however, at the limit to the danger of deadlock.

With $\Delta_1(s) = \Delta(s)$, i.e., in the scheme of Fig. 5.1 the nonlinear loop cannot be stabilized.

With

$$\Delta_1(s) = \begin{bmatrix} s^3 + 4.28024s^2 + 0.176102s + 1.92587 \\ s^2 + 1.92156s - 1.06993 \\ -2s + 3.11627 \end{bmatrix}$$

$$\begin{matrix} -1.31427s^2 - 8.01970s - 8.25496 \\ s^2 + 5.27061s + 4.58609 \\ 1.90820 \end{matrix}$$

$$\left. \begin{matrix} 2.65400s^2 - 2.25902s + 2.42153 \\ 0.0509125s - 2.05623 \\ s + 8.24944 \end{matrix} \right],$$

however, the transfer matrix $G^1_{Luy}(s)$ satisfies the CC for all $0 < \alpha \le 0.492$. Choosing the fastest possible control, i.e., $\alpha = 0.492$, the compensator is characterized by the polynomial matrices

$$D_C(s) = \begin{bmatrix} s & 0 & 0 \\ 0 & s & 0 \\ 0 & 0 & s \end{bmatrix} \bar{N}(s),$$

and

$$N_C(s) = \begin{bmatrix} 0.492 & 0 & 0 \\ 0 & 0.492 & 0 \\ 0 & 0 & 0.492 \end{bmatrix} \bar{D}(s),$$

and

$$N_{u1}(s) = \begin{bmatrix} -0.28024s^2 - 0.176102s - 1.92587 \\ 2.07844s + 1.06993 \\ -3.11627 \\ 1.31427s^2 - 1.9803s + 8.25496 \\ 0.72939s - 4.58609 \\ -1.9082 \\ -3.654s^2 + 3.25902s - 2.42153 \\ -1.0509125s + 2.05623 \\ -8.24944 \end{bmatrix}.$$

Though the system is at the limit to the danger of deadlock a stabilizing compensator can be designed.

Example 5.9 Consider a system with three inputs and three outputs with real poles at $-1, -1, -1$ and $-2, -2, -2$ and with zeros at $-2 \pm 2j, -4$ as in Example 5.8. However, the static behavior of the system is modified so that the danger of deadlock exists. The system is characterized by its right MFD $G(s) = N(s)D^{-1}(s)$ with

$$N(s) = \begin{bmatrix} s - 0.8 & -0.74 & -0.74 \\ 0 & s+4 & 0 \\ 16 & 0.8 & s+4.8 \end{bmatrix},$$

and the same $D(s)$ as in Example 5.8. The left MFD $G(s) = \bar{D}^{-1}(s)\bar{N}(s)$ is specified by

$$\bar{D}(s) = \begin{bmatrix} 3s^2 + 13.9s + 10.9 & 7.57(s+1) & 1.295(s+1) \\ 3.5(s+1) & 3s^2 + 9.65s + 6.65 & 0.925(s+1) \\ -21(s+1) & -29.1(s+1) & 3s^2 + 3.45s + 0.45 \end{bmatrix} \frac{1}{3},$$

$$\bar{N}(s) = \begin{bmatrix} 3s + 5.5 & 8.35 & -0.925 \\ 0.5 & 3s + 9.65 & 0.925 \\ 21 & -14.7 & 3s + 8.85 \end{bmatrix} \frac{1}{3},$$

and the disturbance input by

$$\bar{N}_d(s) = \begin{bmatrix} 2s + 4.5 & -15 & 16 \\ -4 & 4s - 5 & 1.5 \\ -1 & 17.5 & s - 14.50 \end{bmatrix}.$$

With

$$G(0) = \begin{bmatrix} 0.51 & -0.57 & -0.37 \\ -1 & 1 & 0 \\ 5.8 & 5.4 & 2.4 \end{bmatrix},$$

and

$$G^{-1}(0) = \begin{bmatrix} 0.6 & -0.1575 & 0.0925 \\ 0.6 & 0.8425 & 0.0925 \\ -2.8 & -1.515 & -0.015 \end{bmatrix}$$

the DLC is not satisfied so that deadlock can occur when the compensator is designed on the basis of $\Delta_1(s) = \Delta(s)$. Since the LME is located at $s = -1.27$, only a slight tendency for deadlock exists. Thus, it is not easy to reproduce deadlock in simulations.

With $\Delta_1(s) = \Delta(s)$ and

$$\Delta_r(s) = \begin{bmatrix} s + 0.1 & 0 & 0 \\ 0 & s + 0.1 & 0 \\ 0 & 0 & s + 0.1 \end{bmatrix},$$

the compensator is characterized by

$$D_C(s) = \begin{bmatrix} s & 0 & 0 \\ 0 & s & 0 \\ 0 & 0 & s \end{bmatrix} \bar{N}(s),$$

and

$$N_C(s) = \begin{bmatrix} 0.1 & 0 & 0 \\ 0 & 0.1 & 0 \\ 0 & 0 & 0.1 \end{bmatrix} \bar{D}(s).$$

The vectors ga_i related to the LME have the forms $ga_1 = [1\ 0\ 1\quad 0\ 1\ 1]$ and $ga_2 = [0\ 1\ 0\quad 1\ 0\ 0]$. For the input disturbances $d_i(t) = d_{Si}1(t) - d_{Si}1(t-20), i = 1, 2, 3$ with $d_{S1} = d_{S2} = d_{S3} = 1$, the u_{0is} and y_{0is} are $u_{01s} = 2$, $u_{02s} = 1$, $u_{03s} = 6$ and $y_{01s} = 3.5$, $y_{02s} = 3$, $y_{03s} = 6$.

The vector ga_1 suggests the saturation limits $u_{01} = 2$, $u_{02} = 0.1$, $u_{03} = 6$, $y_{01} = 0.35$, $y_{02} = 3$, and $y_{03} = 6$ (see Sect. 5.5). The simulation enters a deadlock with $u_1(\infty) = 1.5671$, $u_2(\infty) = 0.1$, $u_3(\infty) = -6$ and $y_1(\infty) = 0.35$, $y_2(\infty) = -1.4674$, and $y_3(\infty) = -4.7689$.

Using the saturation limits $u_{01} = 0.2$, $u_{02} = 1$, $u_{03} = 0.6$, $y_{01} = 3.5$, $y_{02} = 0.3$, and $y_{03} = 0.6$ suggested by the vector ga_2, the simulation does not enter a deadlock situation.

When "optimizing" the polynomial matrix $\Delta_1(s)$ (see Remark 2.1) so that the biggest value α can be used in the compensator design, one obtains

$$\Delta_1(s) = \begin{bmatrix} s^2 + 2.031323s + 0.4137457 \\ 0.0098161s + 0.2485035 \\ 7.024741s + 0.6892345 \end{bmatrix}$$

$$\begin{matrix} 2.941777s + 0.3629852 \\ s^2 + 3.248115s + 0.2895524 \\ -4.740932s - 0.8139539 \end{matrix}$$

$$\left. \begin{matrix} -0.4108427s - 0.1480087 \\ 0.0933518s - 0.1179189 \\ s^2 + 3.073995s + 0.5030420 \end{matrix} \right] .$$

With the above $\Delta_1(s)$ the matrix $G_{Luy}^1(s)$ satisfies the CC for all $0 < \alpha \le 0.03$. But even with the biggest value $\alpha = 0.03$ the disturbance rejection is quite poor and the transients only settle after about 200 s.

This example demonstrates again why one should apply a stabilizing control in the presence of input and output constraints only to systems satisfying the DLC.

Unfortunately, the DLC is only sufficient for systems with $p < 4$ inputs and outputs. For systems with $p > 3$, there is the possibility that the danger of deadlock exists even though condition (5.36) is satisfied. This will be demonstrated by the next example.

Example 5.10 Consider a system with four inputs and four outputs whose transfer matrix is characterized by the polynomial matrices

$$N(s) = \begin{bmatrix} 2.6 & 4.4 & 3.2 & 2.2 \\ 0.8 & 2 & 3 & 2 \\ -0.5 & 0.7 & 4.1 & 5.6 \\ 0.3 & -0.3 & 1.1 & 1.6 \end{bmatrix},$$

$$D(s) = \begin{bmatrix} s+1 & 1 & 1 & 0 \\ 0 & s+1 & 1 & 1 \\ 0 & 0 & s+1 & 1 \\ 0 & 0 & 0 & s+1 \end{bmatrix},$$

$$\bar{D}(s) = \begin{bmatrix} 16.14s + 15.348 & 15.68 & 18.02 & -4.916 \\ 8.436 & 16.14s - 1.54 & 16.82 & 2.068 \\ 20.388 & -55.49 & 16.14s + 39.07 & 9.494 \\ 5.244 & -11.79 & 5.73 & 16.14s + 11.682 \end{bmatrix} \frac{1}{16.14},$$

and $\bar{N}(s) = N(s)$. There are four step-like external disturbances which attack via

$$\bar{N}_d(s) = \begin{bmatrix} 15 & 10 & 14 & 12 \\ 9 & 3 & 12 & 6 \\ 2 & 6 & 8 & 4 \\ 3 & 2 & 4 & 2 \end{bmatrix}.$$

The maximum difference degree of all outputs is $\kappa = 1$ so that the parametrizing matrix of the compensating controller has the form

$$\Delta_r(s) = \begin{bmatrix} s+\alpha & 0 & 0 & 0 \\ 0 & s+\alpha & 0 & 0 \\ 0 & 0 & s+\alpha & 0 \\ 0 & 0 & 0 & s+\alpha \end{bmatrix}.$$

With

$$G(0) = \begin{bmatrix} 2.6 & 1.8 & -1.2 & 1.6 \\ 0.8 & 1.2 & 1 & -0.2 \\ -0.5 & 1.2 & 3.4 & 1 \\ 0.3 & -0.6 & 1.4 & 0.8 \end{bmatrix},$$

and

$$G(0)^{-1} = \begin{bmatrix} 0.72 & 10.2 & -6.6 & 9.36 \\ 2.1 & 0.16 & 4.96 & -10.36 \\ -2.1 & 5.22 & 0.42 & 4.98 \\ 4.98 & -12.84 & 5.46 & 0.18 \end{bmatrix} \dfrac{1}{16.14}$$

condition (5.36) of the DLC is satisfied. The LME, however, is located at -1.325, so that the danger of deadlock exists.

The vectors ga_1 and ga_2 related to the LME are $ga_1 = [1\ 0\ 1\ 0 \quad 1\ 0\ 1\ 0]$ and $ga_2 = [0\ 1\ 0\ 1 \quad 0\ 1\ 0\ 1]$.

When applying the disturbance inputs $d_i(t) = d_{Si}1(t) - d_{Si}1(t-10)$, $i = 1, 2, 3, 4$ with $d_{S1} = d_{S3} = -1$, $d_{S2} = d_{S4} = 1$ the saturation limits u_{0is} and y_{0is} are $u_{01s} = 2.5$, $u_{02s} = 4$, $u_{03s} = 9$, $u_{04s} = 6$, and $y_{01s} = y_{04s} = 4$, $y_{02s} = y_{03s} = 3$.

If the compensator is designed for $\alpha = 1$, its transfer behavior is characterized by

$$N_C(s) = \bar{D}(s) \quad \text{and} \quad D_C(s) = \text{diag}[s]\bar{N}(s).$$

Using the saturation limits suggested by ga_1, namely, $u_{01} = 2.5$, $u_{02} = 0.4$, $u_{03} = 9$, $u_{04} = 0.6$ and $y_{01} = 4$, $y_{02} = 0.3$, $y_{03} = 3$, $y_{04} = 0.4$ and applying the above disturbances the inputs remain at $u_1(\infty) = 2.5$, $u_2(\infty) = -0.4$, $u_3(\infty) = 2.1048$, and $u_4(\infty) = -0.6$ and the outputs at $y_1(\infty) = 2.2943$, $y_2(\infty) = 0.3$, $y_3(\infty) = 3$, and $y_4(\infty) = 0.4$.

Using the saturation limits suggested by ga_2 instead, namely, $u_{01} = 0.25$, $u_{02} = 4$, $u_{03} = 0.9$, $u_{04} = 6$ and $y_{01} = 0.4$, $y_{02} = 3$, $y_{03} = 0.3$, $y_{04} = 4$ deadlock also occurs. Now the inputs remain at $u_1(\infty) = 0.25$, $u_2(\infty) = -4$, $u_3(\infty) = 0.9$, and $u_4(\infty) =$

−6 and the outputs at $y_1(\infty) = -0.4$, $y_2(\infty) = -2.5$, $y_3(\infty) = -0.3$, and $y_4(\infty) = -1.065$, i.e., $y_2(t)$ and $y_4(t)$ are not saturated.

With an appropriate polynomial matrix $\Delta_1(s)$, one can obtain a stabilizing control. The best solution is

$$\Delta_1(s) = \begin{bmatrix} 2.6s - 1.033213 & 4.4s + 1.934018 \\ 0.8s - 0.6398567 & 2s + 0.7507747 \\ -0.5s - 1.824907 & 0.7s + 1.051006 \\ 0.3s + 0.5735408 & -0.3s - 1.321264 \end{bmatrix}$$

$$\begin{bmatrix} 3.2s - 0.6011607 & 2.2s + 1.016336 \\ 3s + 0.2100528 & 2s + 1.201720 \\ 4.1s - 0.0765597 & 5.6s + 4.105906 \\ 1.1s + 1.379116 & 1.6s + 0.7774935 \end{bmatrix}.$$

With $\Delta_1(s)$ the transfer matrix $G^1_{Luy}(s)$ satisfies the CC for all $0 < \alpha \le 0.155$. For $\alpha = 0.155$, the transfer matrix of the nominal compensator is specified by

$$N_C(s) = \text{diag}[0.155]\bar{D}(s) \quad \text{and} \quad D_C(s) = \text{diag}[s]\bar{N}(s).$$

Figure 5.3 shows the confined signals in the loop in Fig. 5.2 resulting from the above disturbance inputs with saturation limits $u_{01} = u_{02} = 2$, $u_{03} = u_{04} = 1.5$ and $y_{01} = 3$, $y_{02} = 2.5$, $y_{03} = 2$, $y_{04} = 1.5$. The full lines correspond to the subscript 1, the

Fig. 5.3 Disturbance transients of the nonlinear loop

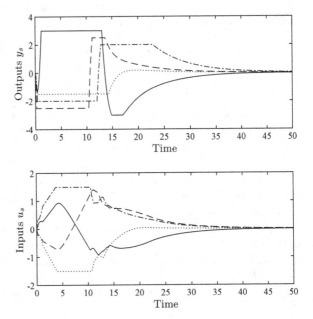

broken lines to the subscript 2, the dash-dotted lines to the subscript 3, and the dotted lines to the subscript 4. The results are not very satisfactory because the dynamics of the loop are quite slow. This is a consequence of the fact that the system has a tendency for deadlock. Though this tendency is relatively slight (see Remark 5.3) it restricts the achievable dynamics.

The final example demonstrates the improvement possible when no danger of deadlock exists.

Example 5.11 Finally, consider a system with D(s) as in Example 5.10 but with a modified numerator matrix

$$
N(s) = \begin{bmatrix} 1 & 0 & 1 & -1 \\ 1 & 2 & 3 & 3 \\ -1 & 0 & 2 & 2 \\ -1 & -2 & -1 & 1 \end{bmatrix}.
$$

The left coprime MFD is now characterized by

$$
\bar{D}(s) = \begin{bmatrix} s+0.7 & 0.2 & 0.2 & -0.3 \\ -0.3 & s+2.2 & 0.2 & 0.7 \\ -0.3 & 0.7 & s+0.2 & 1.2 \\ -0.1 & -0.6 & -0.6 & s+0.9 \end{bmatrix},
$$

and $\bar{N}(s) = N(s)$. The input matrix $\bar{N}_d(s)$ is the same as in Example 5.10. This system also satisfies condition (5.36) of the DLC but now the LME is not located left of -1 so that no danger of deadlock exists.

The loop in Fig. 5.1 does not assure a stable behavior for any parameter α in (5.22).

Looking for the best polynomial matrix $\Delta_1(s)$, one obtains

$$
\Delta_1(s) = \begin{bmatrix} s+1.202465 & -0.4130245 \\ s+1.305889 & 2s+1.480873 \\ -s-0.5436689 & 0.09897267 \\ -s-0.7286837 & -2s-1.478457 \end{bmatrix}
$$

$$
\begin{bmatrix} s+1.165315 & -s-0.6724324 \\ 3s+2.564757 & 3s+2.257121 \\ 2s+1.476806 & 2s+1.385116 \\ -s-0.7892308 & s+0.5709896 \end{bmatrix}.
$$

This matrix assures a stable behavior for all $0 < \alpha \le 0.772$. Choosing $\alpha = 0.772$ the linear compensator is characterized by its polynomial matrices $N_C(s) = \mathrm{diag}[0.772]\bar{D}(s)$ and $D_C(s) = \mathrm{diag}[s]\bar{N}(s)$.

With the same inputs and saturation limits as in Example 5.10 the confined inputs and outputs have the forms shown in Fig. 5.4. The line style distribution is the same

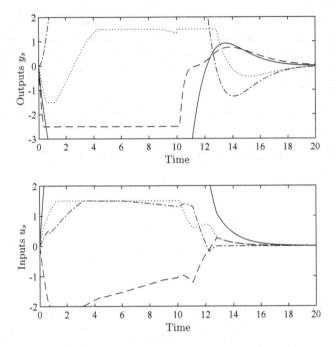

Fig. 5.4 Disturbance transients of the nonlinear loop

as in Fig. 5.3. Compared to the results in Fig. 5.3 the disturbance transients in Fig. 5.4 are much better. This is a consequence of the fact that the system in this example has no tendency for deadlock.

References

1. P. Hippe, J. Deutscher, *Design of Observer-based Compensators—From the Time to the Frequency Domain* (Springer, Berlin, Heidelberg, New York, London, 2009)
2. P. Hippe, *Windup in Control—Its Effects and Their Prevention* (Springer, Berlin, Heidelberg, New York, London, 2006)
3. P. Hippe, C. Wurmthaler, Controller and plant windup prevention in MIMO loops with input saturation, in *Proceedings of the 4th European Control Conference, ECC'97*, Brussels, Belgium (1997), pp. 1891–1896

Appendix

A.1 Nonlinear Model-based Reference Shaping Filter

Using an appropriate model-based trajectory planning reference signals can always be applied to a system with input and output restrictions such that windup effects do not occur. In [1], Sect. 5.4, a nonlinear model-based reference shaping filter was introduced which, given arbitrary external reference inputs $r(t)$, yields reference trajectories $y_M(t)$ and corresponding input signals $u_M(t)$ such that, when $u_M(t)$ is applied to the input of the system in steady state and in the absence of disturbances its output is $y_M(t)$. Furthermore, $u_M(t)$ can be made to stay within desired amplitude limits.

Given a model,

$$\dot{x}_M(t) = A_M x_M(t) + B_M u_M(t), \tag{A.1}$$

$$y_M(t) = C_M x_M(t), \tag{A.2}$$

of the plant with state $x_M \in \mathbb{R}^n$, input $u_M \in \mathbb{R}^p$, and output $y_M \in \mathbb{R}^m$. For simplicity, we assume $m = p$ and that the system (A.1), (A.2) is completely controllable and observable.

Figure A.1 depicts the block diagram of the nonlinear model-based reference shaping filter.

It incorporates a saturating element $u_{as} = \text{sat}_{u_{0f}}(u_a)$ whose elements are defined by

$$\text{sat}_{u_{0f}}(u_{ai}) := \begin{cases} u_{0fi} & \text{if } u_{ai} > u_{0fi} > 0 \\ u_{ai} & \text{if } -u_{0fi} \leq u_{ai} \leq u_{0fi}; \quad i = 1, 2, \ldots, p. \\ -u_{0fi} & \text{if } u_{ai} < -u_{0fi} \end{cases} \tag{A.3}$$

If the state feedback in

Fig. A.1 Nonlinear model-based reference shaping filter

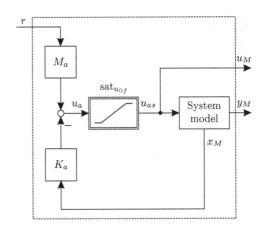

$$u_a(t) = -K_a x_M(t) + M_a r(t), \tag{A.4}$$

where $r \in \mathbb{R}^p$, is designed such that the transfer matrix

$$G_a(s) = K_a(sI - A_M)^{-1} B_M \tag{A.5}$$

in

$$u_a(s) = -G_a(s)u_{as}(s) \tag{A.6}$$

satisfies the Circle Criterion (CC), the reference shaping filter is stable for all reference inputs $r(t)$. Vanishing tracking errors $y_M(\infty) - r(\infty)$ are assured by

$$M_a = [C_M(-A_M + B_M K_a)^{-1} B_M]^{-1}. \tag{A.7}$$

If the saturation limits $u_{0fi}, i = 1, 2, \ldots, p$ are chosen such that $u_{0fi} < u_{0i}, i = 1, 2, \ldots, p$, where the u_{0i} are the saturation limits at the input of the system, no reference input $r(t)$ leads to a saturation at the plant input.

In the sequel, it is shown how this filter can be used to apply reference signals without causing windup in systems with input and output saturation.

A.2 Windup Prevention for Reference Inputs

We assume that the system to be controlled is described by its transfer behavior $y(s) = G(s)u_s(s)$ with

$$G(s) = N(s)D^{-1}(s) = \bar{D}^{-1}(s)\bar{N}(s), \tag{A.8}$$

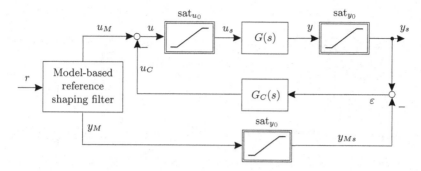

Fig. A.2 Loop with nonlinear reference shaping filter

and that there is an input nonlinearity $u_s = \text{sat}_{u_0}(u)$ whose elements are defined by

$$\text{sat}_{u_0}(u_i) := \begin{cases} u_{0i} & \text{if } u_i > u_{0i} > 0 \\ u_i & \text{if } -u_{0i} \leq u_i \leq u_{0i}; \quad i = 1, 2, \ldots, p. \\ -u_{0i} & \text{if } u_i < -u_{0i} \end{cases} \tag{A.9}$$

The measured output vector $y_s(t) \in \mathbb{R}^p$ is a saturated version of the output y of the system. The components of the nonlinearity $y_s = \text{sat}_{y_0}(y)$ are defined by

$$\text{sat}_{y_0}(y_i) := \begin{cases} y_{0i} & \text{if } y_i > y_{0i} > 0 \\ y_i & \text{if } -y_{0i} \leq y_i \leq y_{0i}; \quad i = 1, 2, \ldots, p. \\ -y_{0i} & \text{if } y_i < -y_{0i} \end{cases} \tag{A.10}$$

The transfer matrix $G_C(s)$ of the compensator in $u_C(s) = G_C(s)\varepsilon(s)$ is represented by

$$G_C(s) = D_C^{-1}(s)N_C(s). \tag{A.11}$$

Figure A.2 shows the loop with the nonlinear model-based reference shaping filter and a model $y_{Ms} = \text{sat}_{y_0}(y_M)$ of the sensor saturation.

Even if the reference signal exceeds the sensor saturation limit (for a short period of time only, of course, because otherwise the loop would be open permanently), the signals y_s and y_{Ms} coincide. Thus, ε vanishes for all reference inputs, and consequently output saturation does not trigger windup for arbitrary reference signals. And since also input saturation does not become active, reference signals do not cause windup in the loop of Fig. A.2.

To prevent controller windup due to input saturation the compensator is often realized in the observer structure (see [1], p. 36 or [2], p. 56), i.e., according to

$$u_C(s) = \Delta^{-1}(s)N_C(s)\varepsilon(s) + \Delta^{-1}(s)N_u(s)u_s(s), \tag{A.12}$$

with

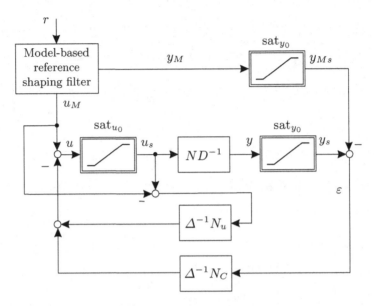

Fig. A.3 Loop structure where reference inputs do not cause windup

$$N_u(s) = D_C(s) - \Delta(s). \tag{A.13}$$

This necessitates a slight modification of the scheme in Fig. A.2.

Figure A.3 shows the scheme assuring that reference signals do not trigger windup effects when the compensator is realized in the observer structure.

Example A.1 As a simple demonstrating example take the SISO system with transfer function

$$G(s) = \frac{s+2}{(s+1)^3}.$$

The compensator with integral action is designed to obtain a very good disturbance rejection. With $\tilde{D}(s) = (s+10)^3$ and $\Delta(s) = (s+2)(s+10)^2$ its transfer function is

$$G_C(s) = \frac{N_C(s)}{D_C(s)} = \frac{856s^3 + 9858s^2 + 49953s + 100000}{s^3 + 49s^2 + 94s},$$

and one obtains

$$N_u(s) = 27s^2 - 46s - 200.$$

It is assumed that there is an input saturation with $u_0 = 1.5$ and a sensor saturation with $y_0 = 1$. For the design of the reference shaping filter, we use a time-domain realization of the system with

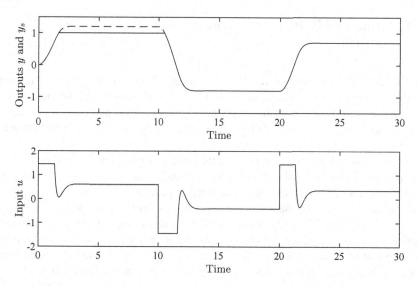

Fig. A.4 Reference transients of the example

$$A_M = \begin{bmatrix} -3 & 1 & 0 \\ -3 & 0 & 1 \\ -1 & 0 & 0 \end{bmatrix}, B_M = \begin{bmatrix} 0 \\ 1 \\ 2 \end{bmatrix}, C_M = \begin{bmatrix} 1 & 0 & 0 \end{bmatrix}.$$

With

$$K_a = \begin{bmatrix} 2 & 2 & 2 \end{bmatrix}$$

the transfer function $G_a(s) = K_a(sI - A_M)^{-1}B_M$ satisfies the CC, i.e., the frequency response $G_a(j\omega)$ stays right of the point $(-1, j0)$ in the complex plane, and (A.7) yields $M_a = 13.5$. With the saturation limit $u_{0f} = 1.45$ the reference shaping filter is completely parametrized.

Figure A.4 shows the reaction of the loop in Fig. A.3 to an external reference signal $r(t) = r_{01}1(t) - r_{02}1(t - 10) + r_{03}1(t - 20)$ with $r_{01} = 1.2$, $r_{02} = 2$ and $r_{03} = 1.5$. In the upper part, the dotted line shows the output y of the system and the full line depicts the output of the sensor. Although the reference signal exceeds the sensor saturation limit during the first 10 s, windup is not triggered because the signal ε vanishes. The lower part of Fig. A.4 shows the input signal to the plant which does not exceed the input saturation limit $u_0 = 1.5$. Thus, no windup is triggered by arbitrary external reference inputs.

A.3 Simplified Reference Tracking (with Compensating Control)

The model-based nonlinear reference shaping filter in Sect. A.1 allows the application of arbitrary external reference inputs $r(t)$ such that windup effects are neither caused by input nor by output constraints. This uses a model of the plant and it generates signals $u_M(t)$ and $y_M(t)$, such that when $u_M(t)$ is the input of the plant its output is $y_M(t)$.

For systems with p inputs and outputs, it is often desirable that the modification of one reference input $r_i(t)$ only affects $y_i(t)$ and none of the remaining $p-1$ outputs, i.e., the reference behavior should be diagonally decoupled. Furthermore, it is desirable that the decoupled behavior is not disrupted by input signal saturation, called the *directionality* problem [3]. Thus, the filter must not only assure a decoupling control but also directionality preservation. An appropriately modified nonlinear model-based reference shaping filter can solve both problems (see [1], Sect. 6.6). But such decoupling control cannot be designed for arbitrary MIMO systems [4].

In the sequel, a simplified approach to reference tracking is presented. It becomes possible when the controller with transfer behavior $G_C(s)$ is designed to be a *compensating* one as described in Sect. 3.3 for SISO and in Sect. 3.5 for MIMO systems. In addition to being easier to design, the simplified filter has the following advantages:

- It uses a very simple system model depending on one parameter β only.
- It has a diagonal structure for MIMO systems and the p models also depend on one parameter β only.
- It allows to design a diagonally decoupled reference behavior also for systems which are not diagonally decouplable by state feedback.

The modified reference shaping filter described in the sequel has the same structure as the one shown in Fig. A.1, however, with input r and output y_M only, as the signal u_M is not required here. The SISO filter uses a simplified model $y_M(s) = G_M(s)u_a(s)$ with transfer function

$$G_M(s) = \frac{\beta^\kappa}{(s+\beta)^\kappa}, \tag{A.14}$$

where κ is the difference degree of the plant and $\beta > 0$. In the sequel, we use a state-space realization (A_M, B_M, C_M) of the model (A.14). If the state feedback

$$u_a(t) = -K_a x_M(t) + M_a r(t) \tag{A.15}$$

is designed such that

$$G_a(s) = K_a(sI - A_M)^{-1} B_M \tag{A.16}$$

in

$$u_a(s) = -G_a(s)u_{as}(s) \tag{A.17}$$

satisfies the CC, the reference shaping filter is stable for all reference inputs $r(t)$. Vanishing tracking errors $y_M(\infty) - r(\infty)$ are assured by

$$M_a = \frac{1}{C_M(-A_M + B_M K_a)^{-1} B_M}. \tag{A.18}$$

When the saturation limit in the filter is chosen as $u_{0f} = y_0$ set points in the whole range of the achievable outputs can be realized.

The realization of the reference tracking is the same as shown in Fig. A.2 where, however, the signal $u_M(t)$ does not exist and the model $y_{Ms} = \text{sat}_{y_0}(y_M)$ is not required, since set points beyond the sensor saturation limits are not considered here. The resulting input signals $u(t)$ in this scheme can, of course, exceed the always-existing input signal limitations.

For a given reference input $r(t)$, the amplitude of $u(t)$ depends on the value α in the compensator design and on the value β, characterizing the dynamics of the model (A.14). Since α has been chosen to obtain a desired amount of disturbance attenuation β can be used to prevent $u(t)$ from exceeding the saturation limit u_0. This is, e.g., possible in simulations with the maximum required reference inputs.

Example A.2 As a demonstrating example consider a simple oscillating system with transfer function

$$G(s) = \frac{10}{s^2 + 2s + 10}.$$

It is assumed that an input saturation with $u_0 = 5$ and an output saturation with $y_0 = 5$ exist, and that the compensating design of Sect. 3.3 is parametrized by $\Delta_r(s) = (s + 10)^2$. This yields the compensator

$$G_C(s) = \frac{100(s^2 + 2s + 10)}{10s^2 + 200s}.$$

Since the difference degree of the system is $\kappa = 2$, the model (A.14) has a state-space realization with, e.g.,

$$A_M = \begin{bmatrix} -2\beta & -\beta^2 \\ 1 & 0 \end{bmatrix}, B_M = \begin{bmatrix} 1 \\ 0 \end{bmatrix}, C_M = \begin{bmatrix} 0 & \beta^2 \end{bmatrix}.$$

When placing the two eigenvalues of the model by the state feedback (A.15) to the location -5.7β the transfer function (A.16) satisfies the CC and K_a becomes $K_a = [9.4\beta \quad 31.49\beta^2]$, and M_a has the form $M_a = 32.49$. For $\beta \leq 6$ and $u_{0f} = 5$, the input signal $u(t)$ does not exceed the saturation limit $u_0 = 5$ for any realizable reference step input.

Figure A.5 shows the responses of the loop in Fig. A.2 for $\beta = 6$ to reference steps $r(t) = r_s 1(t)$ with $r_s = 0, 1, 2, 3, 4, 5$. This β assures unconstrained signals $u(t)$ for all reference inputs.

Fig. A.5 Reference
transients of the example

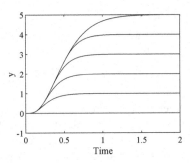

For MIMO systems, the reference shaping filter is as simple as for SISO systems. It
has a diagonal structure with identical diagonal SISO elements of the order κ (see
Eq. (3.33)), with the only exception that the saturation limits $u_{0fi}, i = 1, 2, \ldots, p$
of the p filters can have different values depending on the possibly differing ranges
of the corresponding reference inputs.

Example A.3 As a demonstrating example consider a system with two inputs and
outputs which is not diagonally decouplable by state feedback. Its transfer matrix is
characterized by its MFDs $N(s)D^{-1}(s)$, and $\bar{D}^{-1}(s)\bar{N}(s)$ with

$$
N(s) = \begin{bmatrix} 1 & 1 \\ -1 & s+2 \end{bmatrix}, \quad D(s) = \begin{bmatrix} s^2 + 3s + 1 & s+1 \\ -s-2 & s^2 + 3s + 2 \end{bmatrix},
$$

and

$$
\bar{N}(s) = \begin{bmatrix} 1 & 3 \\ 0 & s+3 \end{bmatrix}, \quad \bar{D}(s) = \begin{bmatrix} s^2 + 2s - 1 & 2s + 4 \\ -s-2 & s^2 + 4s + 4 \end{bmatrix}.
$$

It is assumed that there are saturating sensors with $y_{01} = 4$ and $y_{02} = 5$ and input
constraints with $u_{01} = u_{02} = 10$. The maximum difference degree κ of all outputs
is 2. Therefore, the compensator of Sect. 3.5 is parametrized by

$$
\Delta_r(s) = \begin{bmatrix} (s+\alpha)^2 & 0 \\ 0 & (s+\alpha)^2 \end{bmatrix}.
$$

For $\alpha = 5$, the transfer matrix $G_C(s) = D_C^{-1}(s)N_C(s)$ of the compensator with integral action is characterized by

$$
N_C(s) = \begin{bmatrix} 25 & 0 \\ 0 & 25 \end{bmatrix} \bar{D}(s), \quad \text{and} \quad D_C(s) = \begin{bmatrix} s^2 + 10s & 0 \\ 0 & s^2 + 10s \end{bmatrix} \bar{N}(s).
$$

With $\kappa = 2$ the models $G_{Mi}(s), i = 1, 2$ have the form of $G_M(s)$ in Example A.2, and
this is also true for the state-space representation and the parameters $K_{a1} = K_{a2}$ and
$M_{a1} = M_{a2}$. The only difference in the two diagonal elements is the saturation limits
u_{0fi} which are assumed to be $u_{0f1} = 4$ and $u_{0f2} = 5$. If reference step inputs starting

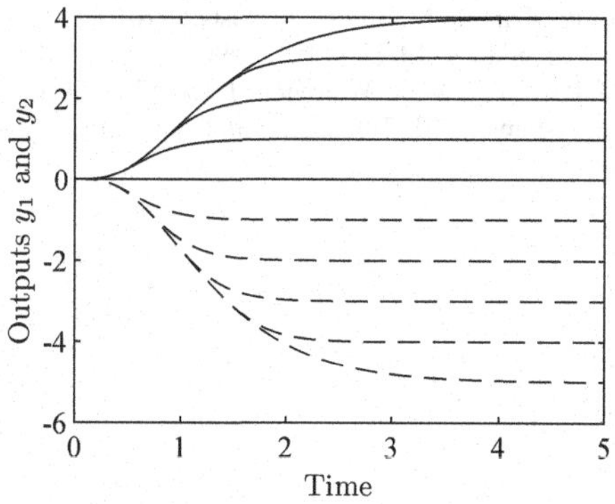

Fig. A.6 Reference
transients of the example

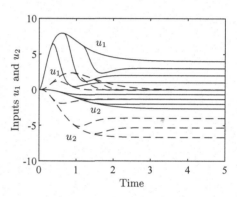

Fig. A.7 Corresponding
plant input signals

from zero must be applied, the inputs $u_i(t)$ in the loop of Fig. A.2 stay within their
unconstrained range for $\beta \leq 2$. With $\beta = 2$ the filter is completely parametrized.

Figure A.6 shows the reactions of the loop in Fig. A.2 to reference inputs $r_i(t) =$
$r_{Si}1(t)$, $i = 1, 2$. The full lines depict the reactions to $r_{S1} = 1, 2, 3, 4$ with $r_{S2} = 0$
and the broken lines the reactions to $r_{S1} = 0$ and $-r_{S2} = 1, 2, 3, 4, 5$. Though the
system is not diagonally decouplable by state feedback, the reference behavior is
clearly decoupled and there are no directionality problems.

Figure A.7 shows the corresponding input signals $u_1(t)$ and $u_2(t)$.

When both reference inputs are applied with opposite signs the input signal amplitudes approach the saturation limits $u_{01} = u_{02} = 10$.

References

1. P. Hippe, *Windup in Control—Its Effects and Their Prevention* (Springer, Berlin, Heidelberg,
   New York, London, 2006)
2. P. Hippe, J. Deutscher, *Design of Observer-based Compensators—From the Time to the Frequency Domain* (Springer, Berlin, Heidelberg, New York, London, 2009)

3. P.J. Campo, M. Morari, Robust control of processes subject to saturation nonlinearities. Comput. Chem. Eng. **14**, 343–358 (1990)
4. P.L. Falb, W.A. Wolovich, Decoupling in the design and synthesis of multivariable control systems. IEEE Trans. Autom. Control **12**, 651–659 (1967)

Printed in the United States
by Baker & Taylor Publisher Services